Advancing Maths for AQA
STATISTICS SS2/SS3

Roger Williamson and Gill Buqué

Series editors
**Roger Williamson   Sam Boardman
Ted Graham   David Pearson**

heinemann.co.uk
✓ Free online support
✓ Useful weblinks
✓ 24 hour online ordering

01865 888058

Heinemann
Inspiring generations

Heinemann is an imprint of Pearson Education Limited,
a company incorporated in England and Wales, having
its registered office at Edinburgh Gate, Harlow, Essex, CM20 2JE.
Registered company number: 872828

Heinemann is a registered trademark
of Pearson Education Limited

First published in 2004

ARP Impression 98

British Library Cataloguing in Publication Data is available from the British
Library on request.

ISBN 978 0 435 51340 5

Edited by Alex Sharpe, Standard Eight Limited
Typeset and illustrated by Tech-Set Limited, Gateshead, Tyne & Wear
Original illustrations © Harcourt Education Limited, 2004
Cover design by Miller, Craig and Cocking Ltd
Printed in Great Britain by Clays Ltd, St Ives plc

Acknowledgements
The publishers and authors' thanks are due to the AQA for permission to
reproduce questions from past examination papers.

The answers have been provided by the authors and are not the responsibility
of the examining board.

Every effort has been made to contact copyright holders of material reproduced
in this book. Any omissions will be rectified in subsequent printings if notice is
given to the publishers

About this book

This book is one in a series of textbooks designed to provide you with exceptional preparation for AQA's 2004 Mathematics Specification B. The series authors are all senior members of the examining team and have prepared the textbooks specifically to support you in studying this course.

Finding your way around

The following are there to help you find your way around when you are studying and revising:
- **edge marks** (shown on the front page) – these help you to get to the right chapter quickly;
- **contents list** – this identifies the individual sections dealing with key syllabus concepts so that you can go straight to the areas that you are looking for;
- **index** – a number in bold type indicates where to find the main entry for that topic.

Key points

Key points are not only summarised at the end of each chapter but are also boxed and highlighted within the text like this:

> Time series are analysed so that they may be projected into the future to make forecasts.

Exercises and exam questions

Worked examples and carefully graded questions familiarise you with the specification and bring you up to exam standard. Each book contains:
- Worked examples and worked exam questions to show you how to tackle typical questions; examiner's tips will also provide guidance;
- Graded exercises, gradually increasing in difficulty up to exam-level questions, which are marked by an [A];
- Test-yourself sections for each chapter so that you can check your understanding of the key aspects of that chapter and identify any sections that you should review;
- Answers to the questions are included at the end of the book.

Contents

CHAPTER 1

Time series analysis

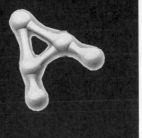

Learning objectives

After studying this chapter, you should be able to:

- understand the concepts of trend, seasonal variation, short-term variation and random variation
- describe a trend
- use moving averages to estimate seasonal effects
- make forecasts by extrapolating the trend and, where appropriate, applying a seasonal effect
- modify forecasts, where appropriate, to allow for short-term variation
- understand that forecasts are merely projections of past patterns and should be treated with caution.

1.1 Introduction

As the name implies a time series is the result of recording a variable at (preferably regular) intervals of time.

For example, daily maximum temperatures, weekly takings of a corner shop or annual profits of a large company can be recorded. The following time series is the annual number of marriages in the United Kingdom in thousands.

Year	1987	1988	1989	1990	1991	1992	1993	1994	1995	1996	1997
Marriages in UK	398	394	392	375	350	356	342	331	322	318	310

Unlike most other topics in the syllabus there is no attempt to obtain a random sample of observations. Rather you look for patterns in the data.

> Time series are analysed so that they may be projected into the future to make forecasts.

There is probably more interest in time series analysis than in any other branch of statistics. This is because all organisations need to make forecasts. Governments need to forecast the number of children who will require school places in future years, electricity manufacturers need to forecast the demand for electricity and pub managers need to forecast the demand for beer.

Time series analysis is concerned with examining the past behaviour of a time series to see if patterns can be discerned. These patterns can then be projected into the future. Unfortunately this does not mean that we can tell what is going to happen in the future. It only means that we can say what will happen if current patterns continue.

> Patterns often do not continue and it is foolish to make exaggerated claims for the reliability of any method of forecasting.

There are however many reasons why it may be useful to make a forecast based on past data. These include the following.

- Having a useful starting point for further discussion. For example the pub manager may know that past trends indicate that 220 bottles of Sunderland Brown Ale will be needed to meet next week's demand. However the manager also knows that an advertising campaign is about to start and so modifies the forecast to 280 bottles.
- Having a useful safeguard against overoptimistic (or overpessimistic) forecasts. A sales manager's forecast which is well in excess of current patterns will need to be closely scrutinised.
- A large supermarket chain will need to forecast the sales of thousands of different items. It will need to have a standard method of making these forecasts.
- Comparing actual results (say sales of chocolate bars) with those predicted from past patterns will enable changes in the pattern, and thus the need for a modified method of forecasting, to be identified.
- It is useful for setting targets. A second division football club may aim to attract more spectators to home matches than would be expected from current trends.
- It may draw attention to the fact that some trends are certain to change and the only question is how. For example it is impossible for the world population to continue to increase at its current rate. If it does we will all be standing on each other's toes. The only question is can human society bring about this change in a humane manner or is it to be left to wars, famines and other horrific events?

Time series may usefully be thought of as being made up of some or all of the following four components:
- trend
- seasonal variation
- short-term non-random variation
- random variation.

These components cannot be identified with certainty but nevertheless provide a useful framework for examining time series.

1.2 Trend

> A trend is a long-term smooth movement.

For example the weekly takings, in £, of a greengrocer over a 12-week period were:

$$890 \quad 900 \quad 910 \quad 920 \quad 930 \quad 940$$

$$950 \quad 960 \quad 970 \quad 980 \quad 990 \quad 1000$$

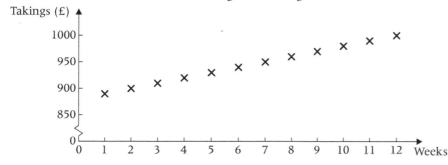

The meaning of long-term and short-term is entirely subjective and depends on the circumstances. A meteorologist studying climatic changes may regard long-term as 1000 years and short-term as 20 years; a speculator on the stock exchange may regard long-term as 6 months and short-term as 2 hours.

The vertical scale does not start at zero. This enables the trend to be seen more clearly. However it does exaggerate the trend. If the vertical scale does not start at zero this should always be indicated on the axis.

The series shows an upward linear trend. It is easy to extend this trend to forecast that next week the takings will be £1010. However there is no guarantee that this will be correct.

It is obvious that the series above is artificial. Real data, such as the UK marriage data shown below, does not behave with such regularity. Indeed the word trend implies that short-term or random deviations from this pattern should be ignored.

Worked example 1.1

(a) Plot a graph of the annual number of marriages in the UK from 1987 to 1997 and describe the trend.

Year	1987	1988	1989	1990	1991	1992	1993	1994	1995	1996	1997
Marriages in UK (thousand)	398	394	392	375	350	356	342	331	322	318	310

Source: Office for National Statistics.

(b) The following table shows marriages in the UK classified by the age of the partners.

Numbers

	1987	1988	1989	1990	1991	1992	1993	1994	1995	1996	1997
Males:											
Under 21 years	24 269	20 608	19 070	15 930	13 271	11 031	8 767	7 091	6 302	5 497	5 126
21–24	118 355	109 482	102 977	92 270	79 877	74 458	65 129	56 877	48 432	42 488	36 875
25–29	119 808	120 939	123 491	122 800	115 637	118 255	114 101	111 108	105 218	101 647	97 345
30–34	51 389	53 865	56 442	56 966	56 970	62 470	63 848	65 490	68 245	69 867	70 904
35–44	48 598	51 329	51 411	49 984	48 147	51 125	50 553	51 310	53 350	56 513	58 292
45–54	19 788	21 544	22 329	21 996	20 915	23 290	23 841	24 136	24 786	26 252	26 472
55 and over	15 730	16 282	16 322	15 464	14 922	15 384	15 369	15 220	14 918	15 250	15 204
Females:											
Under 21 years	68 629	59 284	54 256	45 626	38 305	32 618	26 839	22 903	20 643	18 485	17 254
21–24	140 509	134 122	128 411	119 037	105 505	102 494	93 125	84 171	75 071	66 191	59 549
25–29	90 911	95 338	100 531	103 209	99 851	105 223	104 517	102 803	100 644	99 651	97 932
30–34	36 643	39 680	41 989	42 794	43 617	48 514	49 546	52 359	54 819	57 752	58 589
35–44	36 978	39 534	40 290	38 983	37 582	40 075	40 090	41 213	43 115	45 969	47 267
45–54	15 001	16 570	17 172	16 825	16 473	18 504	18 800	19 280	19 720	21 025	21 038
55 and over	9 260	9 521	9 393	8 936	8 406	8 585	8 691	8 503	8 239	8 441	8 589

Source: Office for National Statistics.

Plot the number of marriages and briefly describe the trend for:

(i) Males aged under 21,

(ii) Males aged 35–44.

Solution

(a)

There is a downward, approximately linear trend. The figure for 1991 is below the trend.

> Pick out the main features. Do not describe every tiny detail of the graph.

(b) (i)

Marriages in the UK – males under 21 years

There is a downward non-linear trend. The rate of reduction is decreasing.

> The early, approximately linear, trend could not continue. If it did there would, by now, be a negative number of marriages from this age group.

(ii)

Marriages in the UK – males 35–44

From 1987 to 1994 the number of marriages fluctuates about a horizontal line (that is, there is no trend). From 1994 there appears to be an upward trend.

> Unlike the younger age groups where the trend is downwards.

EXERCISE 1A

1 The following table shows the number of cinema screens in Great Britain and the total number, in millions, of cinema admissions.

Year	1988	1989	1990	1991	1992	1993	1994	1995	1996	1997	1998
Screens	1117	1177	1331	1544	1547	1591	1619	1620	1738	1886	1975
Admissions	75.2	82.9	78.6	88.9	89.4	99.3	105.9	96.9	118.7	128.2	123.4

Source: Office for National Statistics.

(a) (i) Plot the data for number of screens and briefly describe the trend.

(ii) Plot the data for admissions and briefly describe the trend.

(b) Compare the two trends.

2 The following table shows the number of notifications, in hundreds, of measles and of food poisoning in the United Kingdom.

Year	1988	1989	1990	1991	1992	1993	1994	1995	1996	1997	1998
Measles	906	310	156	117	123	120	235	90	69	48	45
Food poisoning	457	592	597	595	721	767	911	926	949	1056	1050

Source: Annual Abstract of Statistics, 2000.

Plot the data and briefly describe the trend for:

(a) measles,

(b) food poisoning.

3 The following table shows the population, in thousands, of the City of Manchester and of Greater Manchester.

Year	1911	1931	1951	1961	1971	1981	1991	1998
Manchester City	714	766	703	657	554	463	439	430
Greater Manchester	2638	2727	2716	2710	2750	2619	2570	2577

Source: Census and Office for National Statistics.

Plot the data and briefly describe the trend for:

(a) the City of Manchester,

(b) Greater Manchester.

Compare the two trends.

1.3 Seasonal variation

This is the most readily understood component of a time series. We all expect more electricity to be used in winter than in summer and more ice creams to be sold in summer than in winter. Seasonal effects do not necessarily refer to the seasons of the year. For example a greengrocer's takings are likely to be higher on Friday and Saturday than on Monday. This is also called a seasonal effect.

A seasonal effect is a regular predictable pattern.

A market stall is open on Tuesday, Friday and Saturday each week. The takings, in £, over a 4-week period are:

Tue	Fri	Sat	Tue	Fri	Sat
320	525	580	335	540	595

Tue	Fri	Sat	Tue	Fri	Sat
350	555	610	365	570	625

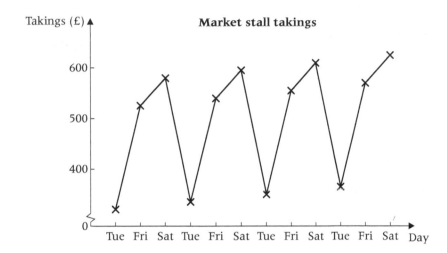

Joining the points with straight lines makes the seasonal pattern clearer.

Don't forget to indicate that the vertical scale does not start at zero.

The graph shows that on Tuesdays takings are low and on Fridays and Saturdays they are high. In other words there is a seasonal effect. To analyse the series it is useful to attempt to remove the seasonal effect (called deseasonalising) and analyse what remains of the series. If a forecast is to be made it will be made from the deseasonalised data and the seasonal effect will then be added back in.

There are many different ways of deseasonalising data and as with many things in statistics it is not possible to say one is right and another is wrong. However the most straightforward way is to use moving averages. In the example above the seasonal effect occurs over the 3 days in a week on which the market stall is open. That is the seasonal pattern repeats itself every three observations. For this reason the observations are compared with a three-point average. The mean of the first three observations is $\frac{(320 + 525 + 580)}{3} = 475$. This should be compared with the middle of the three observations used, that is, with 525. This observation is 50 above the moving average. The next moving average is $\frac{(525 + 580 + 335)}{3} = 480$. This should be compared with 580. This observation is 100 above the moving average.

If the first moving average was compared with, say, the observation for the first Tuesday, it would not be possible to tell whether a difference was due to a seasonal effect or due to a trend.

Day	Tue	Fri	Sat	Tue	Fri	Sat	Tue	Fri	Sat	Tue	Fri	Sat
Takings	320	525	580	335	540	595	350	555	610	365	570	625
Moving average		475	480	485	490	495	500	505	510	515	520	

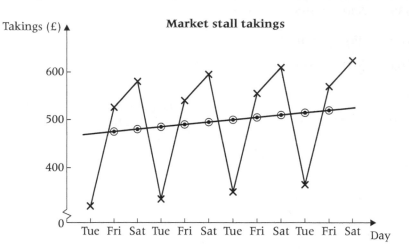

Takings (£)

Market stall takings

Notice that it is not possible to calculate a moving average corresponding to the first Tuesday or the last Saturday.

The moving averages are shown as ◉

The seasonal effect may be removed by calculating a suitable moving average.

The moving average shows an upward linear trend. It is easy to extend this trend and predict that for the next week the moving average will be:

Tue	Fri	Sat
530	535	540

To predict the actual takings for the next week a numerical value of the seasonal effect is required. The following table shows the actual observations minus the moving average.

Day	Tue	Fri	Sat	Tue	Fri	Sat	Tue	Fri	Sat	Tue	Fri	Sat
Takings	320	525	580	335	540	595	350	555	610	365	570	625
Moving average		475	480	485	490	495	500	505	510	515	520	
Takings – moving average		50	100	−150	50	100	−150	50	100	−150	50	

The sign is important.

On Fridays the takings are always 50 above the moving average, on Saturdays they are 100 above the moving average and on Tuesdays they are 150 below the moving average.

The predictions for next week are therefore:

Tuesday	$530 - 150 = 380$
Friday	$535 + 50 = 585$
Saturday	$540 + 100 = 640.$

It is easy to make a mistake and subtract the seasonal effect when you should have added it. If you do make this mistake it will be obvious because your predictions will not follow the earlier pattern of low on Tuesday, high on Friday and Saturday.

Real data

The time series above was clearly fictitious. Real data would not behave with such regularity. Here are some real data. The table shows the daily number of requests, in thousands, for pages from a government department's website over a 3-week period.

Sunday	84	Thursday	151
Monday	162	Friday	178
Tuesday	192	Saturday	74
Wednesday	189	Sunday	84
Thursday	171	Monday	176
Friday	169	Tuesday	149
Saturday	76	Wednesday	181
Sunday	86	Thursday	192
Monday	156	Friday	171
Tuesday	194	Saturday	78
Wednesday	190		

There is clearly a pattern to the weekly requests with the most notable feature being the reduction in requests on Saturdays and Sundays. Since the pattern recurs every 7 days the observations are compared with a seven-point moving average.

The first moving average will be compared with the fourth point.

Day	Requests (thousands)	Moving average (MA)	Requests – MA
Sunday	84		
Monday	162		
Tuesday	192		
Wednesday	189	149.0	40.0
Thursday	171	149.3	21.7
Friday	169	148.4	20.6
Saturday	76	148.7	−72.7
Sunday	86	148.9	−62.9
Monday	156	146.0	10.0
Tuesday	194	147.3	46.7
Wednesday	190	147.0	43.0
Thursday	151	146.7	4.3
Friday	178	149.6	28.4
Saturday	74	143.1	−69.1
Sunday	84	141.9	−57.9
Monday	176	147.7	28.3
Tuesday	149	146.7	2.3
Wednesday	181	147.3	33.7
Thursday	192		
Friday	171		
Saturday	78		

It is not possible to calculate a seven-point moving average corresponding to the first three or the last three points.

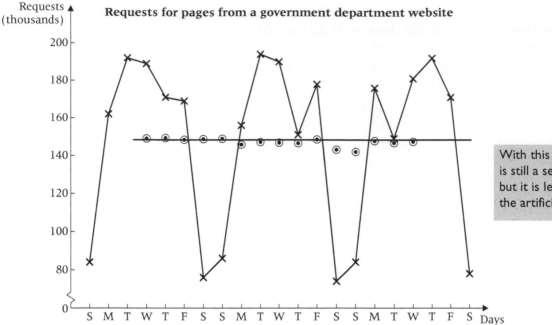

Requests for pages from a government department website

With this real data there is still a seasonal pattern but it is less regular than the artificial data.

From the graph you can see that the moving average is very close to horizontal – in other words there is no trend.

However if you look at the difference between the actual number of requests on, say, Wednesdays and the moving average you see that it is 40.0, 43.0 and 33.7. These three numbers are not the same but they are similar. The best estimate that can be made for the seasonal effect for Wednesdays is the mean of these three numbers.

That is $\dfrac{(40.0 + 43.0 + 33.7)}{3} = 38.9$

For the other days of the week only two moving averages are available and so the best that can be done is to average the (signed) deviation from the moving average for these 2 days.

	Mon	Tue	Wed	Thu	Fri	Sat	Sun
			40.0	21.7	20.6	−72.7	−62.9
	10.0	46.7	43.0	4.3	28.4	−69.1	−57.9
	28.3	2.3	33.7				
Estimated seasonal effect	19.1	24.5	38.9	13.0	24.5	−70.9	−60.4

The sign is important.

This is the average of the numbers in the column above. If there was more data we could improve the estimate.

The graph indicates that there is no noticeable trend in the moving average and so to forecast the number of requests for the next week you could start with the current value of the moving average of 147.3 and apply the estimated seasonal effects. Monday has been estimated to be 19.1 above the moving average and so our forecast for Monday would be:

$$147.3 + 19.1 = 166$$

It is reasonable to give three significant figures in this case. If nothing happens to disturb the current pattern the past data suggests that you cannot hope for the forecasts to be accurate to 3 s.f. You could hope for most of them to be accurate to 2 s.f. To give more than 3 s.f. suggests that you are claiming unattainable levels of accuracy for your forecasts.

> The seasonal effects could be adjusted to make their mean zero. This will make little difference and is not essential.

To complete the forecasts for the week:

Tuesday	$147.3 + 24.5 = 172$
Wednesday	$147.3 + 38.9 = 186$
Thursday	$147.3 + 13.0 = 160$
Friday	$147.3 + 24.5 = 172$
Saturday	$147.3 - 70.9 = 76$
Sunday	$147.3 - 60.4 = 87$

> Don't forget to compare these forecasts with the given data to check you have applied the seasonal effect correctly.

Worked example 1.2

A market stall holder has been selling petfood on Thursday, Friday and Saturday each week for several years. The stall holder employs a consultant to analyse the business. The consultant collects the following data of takings for the last 3 weeks.

Day	1	2	3	4	5	6	7	8	9
	Thu	Fri	Sat	Thu	Fri	Sat	Thu	Fri	Sat
Takings (£)	278	396	592	312	409	622	315	431	621

(a) Plot the data together with a suitable moving average. [A]

(b) Use regression analysis to predict the value of the moving average for the Saturday of the next week. [A]

> Regression analysis is discussed in S1 Chapter 8.

(c) Based on your answer in **(b)** predict the takings for the Saturday of the next week. [A]

(d) The consultant reports that the stall will be taking about £1450 on Saturday in a year's time and this will rise to £2330 on a Saturday in 2 years' time. Comment on the consultant's report.

Solution

(a) Day

Day	1	2	3	4	5	6	7	8	9
	278	396	592	312	409	622	315	431	621
MA		422	433.3	437.7	447.7	448.7	456	455.7	

> A three-point moving average is appropriate.

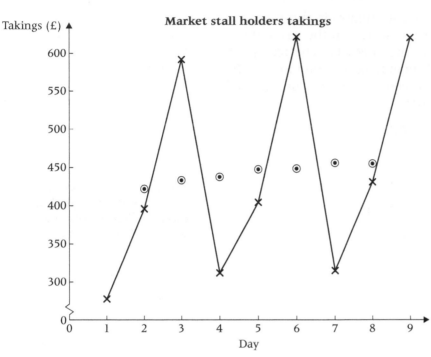

(b) Since the question specifies regression should be used you need to regress the moving average (y) on the number of the day (x). Since the moving averages only run from day 2 to day 8 only these 7 days can be used.

The data can be entered into a calculator to give:

$$y = 415 + 5.62x$$

For the next week Thursday will be day 10, Friday will be day 11 and Saturday will be day 12. The predicted moving average for Saturday of the next week will be:

$$415 + 5.62 \times 12 = 482$$

> Note that in this question the days were numbered for you. If they had not been you would have had to number them yourself. The given numbering is probably the simplest but there is no reason why you cannot start with zero or any other number.

(c) The differences between the takings and the moving averages on the two Saturdays where this can be calculated is:

$$592 - 433.3 = 158.7$$
$$622 - 448.7 = 173.3$$
$$\text{mean} = 166.0.$$

Prediction for Saturday of next week is $482 + 166 = 648$.

> There is no moving average available to compare with the third Saturday.

> Alternatively you could calculate the residuals from the regression line.

> Do not give more than 3 s.f.

(d) The consultant has apparently extrapolated 3 weeks' data to 1 year and 2 years ahead. It is foolish to extrapolate so little data so far ahead. The number of significant figures gives a wholly spurious impression of the likely accuracy of the predictions.

Worked example 1.3

The following table shows the expenditure, in £ million, by UK households on air travel. The figures have been adjusted to constant 1995 prices. These should be used throughout the question.

Year	1997	1998	1999
Quarter 1	1206	1340	1454
Quarter 2	1676	1648	1954
Quarter 3	2139	2123	2530
Quarter 4	1384	1589	1780

Source: Office for National Statistics.

Examine the table carefully. To arrange the observations in order of time you need to go down the first column, then down the second, etc. In other questions you might have to go across rows.

(a) Plot the data together with a suitable moving average.

(b) Describe the trend shown by the moving average.

(c) By extrapolating the trend shown by the last six moving averages predict the moving average for Quarter 1, 2000.

(d) Predict the actual expenditure on air travel, for Quarter 1, 2000.

(e) The actual expenditure in Quarter 1, 2000, was £1646 million. Compare this with your forecast and comment.

(f) Comment briefly on the fact that the data in this question is in constant 1995 prices.

Solution

(a) A four-point moving average is appropriate here. This means that to avoid contaminating the seasonal effect with a possible trend you will need to plot the first moving average halfway between Quarter 1 and Quarter 2.

This will always occur if there are an even number of points in the moving average.

The moving average has been shown on alternate lines to indicate that it corresponds to a point halfway between the two adjoining quarters.

Year		Expenditure £million	Moving average
1997	Q1	1206	
	Q2	1676	
			1601
	Q3	2139	
			1635
	Q4	1384	
			1628
1998	Q1	1340	
			1624
	Q2	1648	
			1675
	Q3	2123	
			1703
	Q4	1589	
			1780
1999	Q1	1454	
			1882
	Q2	1954	
			1929
	Q3	2530	
	Q4	1780	

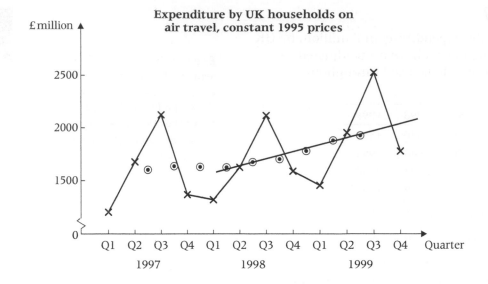

Expenditure by UK households on air travel, constant 1995 prices

(b) The moving average starts off horizontally (i.e. there is no trend) but then shows a small, approximately linear, upward trend.

(c) Moving average for Quarter 1, 2000 about £2100 million.

(d) You need to compare the Quarter 1 figures for 1998 and 1999 with the moving average. This cannot be done directly from the table. It is perfectly acceptable to do this graphically. It can also be done numerically by taking the mean of the two adjoining moving averages.

The moving average immediately before Quarter 1, 1998 is 1628 and the one immediately after is 1624.

The mean is $\dfrac{(1628 + 1624)}{2} = 1626$. This is sometimes called the centred moving average.

The observed value for Quarter 1, 1998 is 1340, which is $1340 - 1626 = -286$ from the moving average.

Making the same calculation for Quarter 1, 1999 gives a centred moving average of $(1780 + 1882)/2 = 1831$. The observed value of 1454 is $1454 - 1831 = -377$ from the moving average.

Taking the mean of -286 and -377 gives -331.5.

The prediction for Quarter 1, 2000 is $2100 - 331.5 = 1770$.

If there are more data or if the seasonal effects for the other quarters are required it is worth putting the calculations in a table.

		Expenditure £million	Moving average	Centred moving average (CMA)	Moving average −CMA
1997	Q1	1206			
	Q2	1676			
			1601		
	Q3	2139			
			1635		
	Q4	1384			
			1628		
1998	Q1	1340		1626	−286
			1624		
	Q2	1648			
			1675		
	Q3	2123			
			1703		
	Q4	1589			
			1780		
1999	Q1	1454		1831	−377
			1882		
	Q2	1954			
			1929		
	Q3	2530			
	Q4	1780			

These are the means of the adjoining moving averages. If you were asked to estimate the seasonal effect for all quarters you would need to complete the last two columns.

(e) The actual value is substantially less than that predicted. Not too much should be read into the figures for one quarter but this suggests that the upward trend of the immediately preceding quarters may have slowed down.

(f) The fact that the data is given at constant 1995 prices means that an upward trend represents a real increase in expenditure on air travel. If the data had not been at constant prices the upward trend might merely reflect inflation.

EXERCISE 1B

1 A mobile fish shop calls at a village on Mondays, Wednesdays and Fridays. The value of the fish sold over a 3-week period was as follows:

Day	M	W	F	M	W	F	M	W	F
Sales, £	17	31	59	20	27	63	21	23	66

Plot the data together with an appropriate moving average.

2 A theatre opens from Tuesday to Saturday. The theatre holds a maximum of 600 people. The attendances for the first 3 weeks of a new play are shown in the table below.

	Week 1					Week 2					Week 3				
T	W	Th	F	S	T	W	Th	F	S	T	W	Th	F	S
199	216	230	320	430	228	209	250	347	470	254	276	296	402	504

(a) Plot a graph of the data together with a suitable moving average.

(b) A severe storm occurred shortly before one of the performances. Which performance do you think this was? Give a reason.

(c) Predict the attendance on the Tuesday of week 4.

(d) Explain why the trend shown by the moving average cannot continue over a long period of time.

3 The quarterly sales (£ thousands) of an agricultural chemical company over a 2-year period are shown below.

	1999				2000		
Q1	Q2	Q3	Q4	Q1	Q2	Q3	Q4
165	170	125	149	x	$1.2x$	130	149

An appropriate moving average is calculated.

(a) Find the value of the first point.

(b) Given that the value of the second point is 160 find the value of the third point.

4 The following table shows the expenditure, in £ million, of UK households on horticultural goods. The figures have been adjusted to constant 1995 prices.

Year	1997	1998	1999
Quarter 1	579	629	739
Quarter 2	857	925	1056
Quarter 3	493	482	632
Quarter 4	411	489	550

Source: Annual Abstract of Statistics, 2000.

(a) Plot the data together with a suitable moving average.

(b) Describe, briefly, the trend shown by the moving average.

(c) Predict the expenditure for Quarter 1, 2000 based on the data above.

(d) The actual expenditure for Quarter 1, 2000 was £791 million. Compare this with the forecast you made in (c) and comment.

(e) Suggest a reason for the seasonal effects you have observed.

5 The expenditure, in £ million, of UK households on sports and toys is shown in the following table. The figures have been adjusted to constant 1995 prices.

Year	Quarter 1	Quarter 2	Quarter 3	Quarter 4
1997	1273	1474	1492	2094
1998	1552	1870	1869	2507
1999	1789	2195	2244	3039

Source: Annual Abstract of Statistics, 2000.

(a) Plot the data together with a suitable moving average.

(b) Describe the trend shown by the moving average.

(c) Use regression analysis to predict the moving average for each quarter of 2000.

(d) Predict the actual expenditure for each quarter of 2000.

(e) The actual expenditure in the first quarter of 2000 was £2080 million. Compare this with your forecast and comment.

(f) Suggest a possible reason for the seasonal effects you have observed in the data.

1.4 Random variation

It is clear from the examples above which involve real data that time series do not behave in an entirely regular or predictable manner. This unpredictable component of the series is called random variation. The series illustrated below exhibits random variation about an upward linear trend.

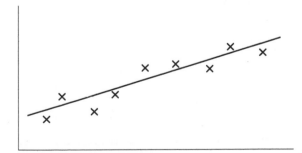

It is impossible to forecast random variation. If it can be forecast then it is not random.

All real time series will contain some random variation which is impossible to forecast.

1.5 Short-term variation

The time series below has a long-term upward linear trend but varies around this trend in a manner which is neither regular nor random.

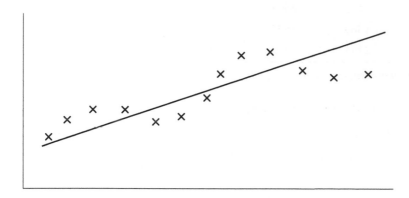

This non-random variation about a trend is called short-term variation.

> Short-term variation is variation about the trend which is neither regular nor random.

Worked example 1.4

Plot points illustrating a time series which displays:

(a) random variation about a downward linear trend,

(b) short-term variation about a downward linear trend,

(c) seasonal variation but no trend,

(d) random variation about a downward non-linear trend.

Solution

(a)

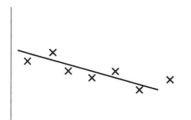

(b)

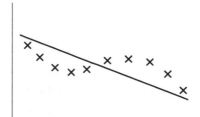

(c)

(d)

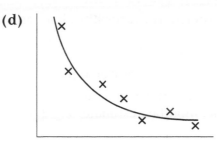

Worked example 1.5

The following data is the annual membership of a rugby club.

Year	1987	1988	1989	1990	1991	1992	1993	1994	1995	1996	1997	1998	1999	2000
Number of members	132	142	149	148	146	144	162	179	196	204	221	233	253	261

(a) Plot the data.

(b) Calculate the equation of the regression line of number of members, y, on year, x, and draw the line on your graph.

(c) Describe the behaviour of the series.

(d) Use the regression line to estimate the number of members of the club in 2001.

(e) Examine your graph and modify the estimate you have made in **(d)**.

Solution

(a)

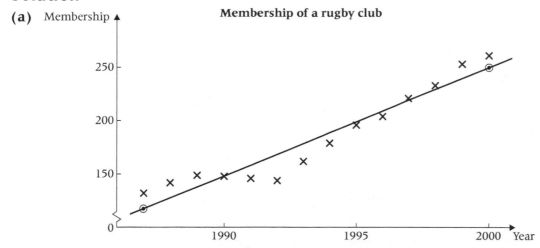

(b) $y = -20\,093.2 + 10.1714x$ from calculator

$x = 1987, \hat{y} = 117.5 \quad x = 2000, \hat{y} = 249.7$

> The raw data has been entered uncoded. If you do this a large number of significant figures are needed because any rounding error in the gradient will be multiplied by about 2000.
>
> Alternatively you can code the data, $1987 \rightarrow 1$, $1988 \rightarrow 2$, etc. before undertaking the calculation.

(c) The series shows short-term variability about an upward linear trend.

(d) $x = 2001$ $\hat{y} = 260$.
Regression line estimates a membership of 260 for 2001.

(e) The graph suggests that the actual number will be above the regression line. Modified estimate, say, 270.

EXERCISE 1C

1 Plot points illustrating a time series which displays:

  (a) random variation about an upward linear trend,

  (b) seasonal variation,

  (c) short-term variation about a downward linear trend,

  (d) random variation about a downward non-linear trend,

  (e) random variation but no trend,

  (f) short-term variation about an upward non-linear trend.

2 **A, B, C, D, E** and **F** are descriptions of time series.

  A random variation about a linear trend

  B seasonal variation about a linear trend

  C random variation about a downward non-linear trend

  D short-term variation about an upward linear trend

  E short-term variation about a linear trend

  F short-term variation about a non-linear trend.

Choose the description **A, B, C, D, E** or **F** which best describes each of the time series below.

(a)
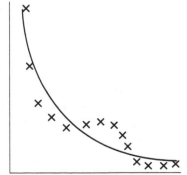

(b)
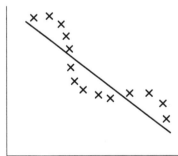

(c)
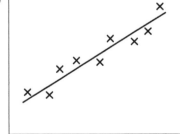

(d)
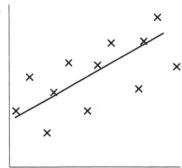

3 The following data is the membership of a lodge of a secretive sexist society. Year 1 is 1986, year 2 is 1987, etc.

Year	1	2	3	4	5	6	7	8	9	10	11	12	13	14	15
Membership	254	239	225	210	206	192	188	172	151	142	120	94	79	59	45

(a) Plot the data.

(b) Calculate the equation of the regression line of membership on year.

(c) Describe the behaviour of the series.

(d) Use the regression line to predict the membership in 2001.

(e) Use your graph to modify your prediction in **(d)**.

(f) What problems would arise in attempting a prediction for 2005?

MIXED EXERCISE

1 The value of goods exported from the United Kingdom to Norway, £ million, is shown in the following table.

Year	1989	1990	1991	1992	1993	1994	1995	1996	1997	1998
Exports	1057	1291	1323	1421	1502	2035	2007	2066	2659	2757

Source: Annual Abstract of Statistics, 2000.

(a) Plot the data.

(b) Describe the trend.

(c) How would the trend be affected if the figures had been adjusted to allow for inflation?

2 The following table shows the number of consultants and the number of nurses and midwives employed in the National Health Service.

Year	1988	1989	1990	1991	1992	1993	1994	1995	1996	1997	1998
Consultants (hundreds)	489	503	518	530	540	556	567	602	622	643	674
Nurses and midwives (thousands)	490	491	487	484	468	446	429	354	356	354	410

Source: Annual Abstract of Statistics, 2000.

Plot the data and describe the trend for:

(a) consultants,

(b) nurses and midwives.

Compare the magnitudes and the trends of the two series.

3 Cecilia takes over a milkround from Alex. She collects money on Tuesdays, Fridays and Saturdays. Her takings for the first 3 weeks were:

Day	T	F	S	T	F	S	T	F	S
Takings (£)	196	340	413	212	373	468	234	400	480

(a) Plot the data together with a suitable moving average.

(b) By extrapolating the trend shown by the moving average predict the moving average for Tuesday, Friday and Saturday of the next week.

(c) Predict the actual takings on Tuesday, Friday and Saturday of the next week.

(d) What reservations would you have about using the method of (b) and (c) to predict the takings for the week, 1 year after she took over the round.

4 The following table gives the oil usage (in tonnes) of a small engineering firm.

Year	Quarter			
	1	2	3	4
1997	125	96	72	119
1998	137	113	88	131
1999	117	118	94	142
2000	162	155	162	176

(a) Plot the data together with a suitable moving average.

(b) During one of the quarters the works was shut down for 3 weeks due to a major breakdown. Which quarter do you think this was? Explain your answer.

(c) During another quarter a leak was found in the oil tank which was then replaced. Which quarter do you think this was. Explain your answer.

5 The following table shows the expenditure, £ million, on stationery of households in the UK. The data has been adjusted to constant 1995 prices.

Year	1997	1998	1999
Quarter 1	810	838	854
Quarter 2	780	764	779
Quarter 3	820	812	814
Quarter 4	1063	1065	1029

Source: Consumer Trends, 2000.

(a) Plot the data together with a suitable moving average.

(b) Use regression analysis to predict the moving average for the four quarters of 2000.

(c) Predict the actual expenditure for the four quarters of 2000.

(d) Given that the actual expenditure for the first quarter of 2000 was £859 million, comment on your method of forecasting.

6 The following data is the weekly output of a weaving mill in terms of lengths of cloth produced. The figures have been adjusted for seasonal effects, including bank holidays.

Week	1	2	3	4	5	6	7	8	9	10	11	12	13	14
Lengths	264	283	298	296	292	287	324	338	393	407	442	466	502	532

(a) Plot the data.

(b) Calculate the equation of the regression line of lengths on week. Draw the line on your graph.

(c) Describe the behaviour of the series.

(d) Use the regression line to estimate the number of lengths produced in week 15.

(e) Examine your graph and modify the estimate you have made in **(d)**.

Key point summary

I Time series are analysed so that they may be projected into the future to make forecasts.   *p1*

2 There are many good reasons for doing this but past patterns do not always continue and so it is unwise to expect the forecasts to be accurate.   *p2*

3 A trend is a long-term smooth movement. It may be estimated by eye from a graph or, if it is approximately linear, by using regression.   *p3*

4 A seasonal effect is a regular predictable pattern, such as cinema audiences being higher on Fridays and Saturdays than on Mondays.   *p6*

5 The seasonal effect may be removed by calculating a suitable moving average. An estimate of the magnitude of the seasonal effect may be made by comparing each point of a time series with the corresponding moving average.   *p8*

6 All real time series will contain some random variation which is impossible to forecast.   *p17*

7 Short-term variation is variation about the trend which is neither regular nor random.   *p18*

Test yourself	What to review
1 Time series may usefully be considered as consisting of some or all of four components. What are these components?	*Section 1.1*
2 What is meant by 'long term' in the context of a time series?	*Section 1.2*
3 You are asked to forecast a series which contains only random variation. What problem would arise?	*Section 1.4*

4 A fishmonger opens 5 days a week from Tuesday to Saturday. The table shows the daily takings, to the nearest £10, over a 2-week period.

Section 1.3

Day	Tu	W	Th	F	S	Tu	W	Th	F	S
Takings (£)	860	540	690	1020	980	900	580	730	1060	1020

Choose a suitable moving average and calculate the values for the two Thursdays.

5 If the trend of the moving average in question **4** is extended it suggests that the moving average for Thursday of the next week would be £898. Forecast the actual takings for Thursday of the next week.	*Section 1.3*
6 What feature of the data in question **4** suggests that it is fictitious?	*Section 1.3*
7 Why is it not possible to calculate a moving average for the second Friday for the data in question **4**?	*Section 1.3*
8 Sketch a time series which shows short-term variation about an upward linear trend.	*Section 1.5*

Test yourself ANSWERS

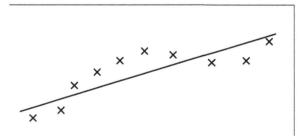

8 This is one possible answer.

7 For a five-point moving average the data for 2 days ahead is necessary. This is not available for the second Friday.

6 The seasonal pattern is exactly the same in both weeks.

5 £770.

4 £818, £858.

3 It is not possible to forecast random variation.

2 Long term is subjective and depends entirely on the context.

1 Trend, seasonal variation, short-term variation, random variation.

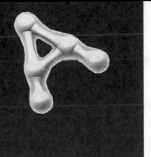

CHAPTER 2
Sampling

Learning objectives

After studying this chapter, you should be able to:

■ define random, stratified, systematic, quota and cluster samples
■ select an appropriate method of sampling for use in particular circumstances
■ understand the advantages and disadvantages of the different methods of sampling.

2.1 Sampling

The purpose of sampling is to obtain information about a population by examining only part of it – a sample.

This may be done for reasons of economy – it is easier and cheaper to test a sample of components than to test the whole batch. It may be done because it is impossible, in practice, to examine the whole population – for example you would never be able to interview every adult in the United Kingdom. Or it may be done because the test is destructive – it would be ridiculous to find the average working life of a large batch of electric light bulbs by testing them all before selling them.

If a sample is to be used to make inferences about a population or to estimate population parameters it is essential that the sample is chosen, as far as possible, to be representative of the population and to avoid bias.

2.2 Random sampling

When you are selecting a sample you need to avoid **bias** – anything which makes the sample unrepresentative. For example, if you want to estimate how often residents of Manchester visit the cinema in a year it would be foolish to stand outside a cinema as the audience is coming out and ask people as they pass. This would give a biased sample as all the people you ask would have been to the cinema at least once that year. You can avoid bias by taking a random sample.

A random sample was defined in S1, Chapter 1.

For a sample to be random every member of the population must have an equal chance of being selected. However, this alone is not sufficient. If the population consists of 10 000 heights and a random sample of size 20 is required then every possible set of 20 heights must have an equal chance of being chosen.

For example, suppose the population consists of the heights of 100 students in a college and you wish to take a sample of size 5. The students' names are arranged in alphabetical order and numbered 00 to 99. A number between 00 and 19 is selected by lottery methods. For example, place 20 equally sized balls numbered 00 to 19 in a bag and ask a blindfolded assistant to pick one out. This student and every 20th one thereafter are chosen and their heights measured. That is if the number 13 is selected then the students numbered 13, 33, 53, 73 and 93 are chosen. Every student would have an equal chance of being chosen. However a sister and brother who were next to each other in the alphabetical list could never both be included in the same sample, so this is **not** a random sample.

> Not every set of five students could be chosen.

> A **random sample** of size n is a sample selected in such a way that all possible samples of size n have an equal chance of being selected.

Usually if you decide to choose five students at random you intend to choose five different students and would not consider choosing the same student twice. This is known as sampling without replacement.

> A random sample chosen without replacement is called a **simple random sample**. If you did allow the possibility of a member of the population being chosen more than once this would be sampling with replacement. A random sample chosen with replacement is called an **unrestricted random sample**.

> Unrestricted random samples are rarely used.

Random numbers

The previous section referred to numbers being selected by lottery methods. In practice it is much more convenient to use random numbers. These are numbers which have been generated so that each digit from 0 to 9 has an equal chance of appearing in each position. They may be obtained from your calculator or from tables. An extract from random number tables is shown on page 27.

> Your calculator may generate random numbers between 0 and 1, say 0.0206. To turn these into random digits ignore 0. and use 0206.

Worked example 2.1

Describe how random numbers could be used to select a simple random sample of size 7 from the 63 residents of Mandela Close who are on the electoral register.

Solution

First number the residents. This is easy as a list of names is already available in the electoral register.

You could place a number by each name, or simply decide that the top name will be 00, the next 01 and so on.

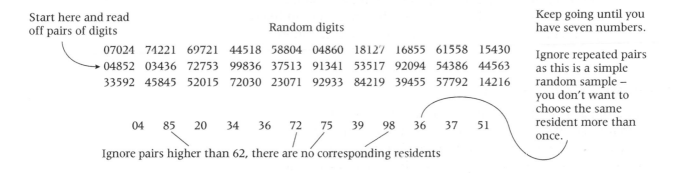

00	Chuzzlewit, M
01	Ngo, S
02	Sodiwala, V
03	Shah, D
04	O'Shea, M

2

Next choose any starting point in the random number tables:

Start here and read off pairs of digits

Random digits

Keep going until you have seven numbers.

```
07024  74221  69721  44518  58804  04860  18127  16855  61558  15430
04852  03436  72753  99836  37513  91341  53517  92094  54386  44563
33592  45845  52015  72030  23071  92933  84219  39455  57792  14216
```

Ignore repeated pairs as this is a simple random sample – you don't want to choose the same resident more than once.

04   85   20   34   36   72   75   39   98   36   37   51

Ignore pairs higher than 62, there are no corresponding residents

When you have seven different numbers in the range 00 to 62, use them to choose the corresponding residents:

You could have numbered from 01 to 63. In this case you would have to ignore 00 and numbers greater than 63.

04   20   34   36   39   37   51

O'Shea

Worked example 2.2

A trade union wished to ask a sample of 100 members to answer a questionnaire about the services it provides. A list of all the 98 650 members of the union is obtained and numbered from 00 000 to 98 649.

Five digit random numbers are read from a table and any numbers over 98 649 are ignored. This continues until 100 five-digit numbers have been obtained. The corresponding union members are contacted with a request to complete the questionnaire.

(a) What is the name given to this method of sampling?

(b) How would this method of sampling be modified if a simple random sample was required?

(c) Which of these two methods of sampling would you recommend?

Solution

(a) This is an **unrestricted random sample**.

(b) The method would be the same except that any repeated random numbers would be ignored. This would prevent the same union member being selected twice.

(c) There is no purpose in asking the same union member to fill in a questionnaire twice and so a simple random sample would be preferred. (However, as the sample is small compared to the population it is, in this case, very unlikely that the same member would have been selected more than once.)

Worked example 2.3

Following a spell of particularly bad weather, an insurance company received 42 claims for storm damage on the same day. Sufficient staff were available to investigate only six of these claims. The others would be paid in full without investigation. The claims were numbered 00 to 41 and the following suggestions were made as to the method used to select the six. In each case six different claims are required, so any repeats would be ignored.

Method 1	Choose the six largest claims.
Method 2	Select two digit random numbers, ignoring any greater than 41. When six have been obtained, choose the corresponding claims.
Method 3	Select two digit random numbers. Divide each one by 42, take the remainder and choose the corresponding claims (e.g. if 44 is selected, claim number 02 would be chosen).
Method 4	As 3, but when selecting the random numbers ignore 84 and over.
Method 5	Select a single digit at random, ignoring 7 and over. Choose this and every seventh claim thereafter (e.g. if 3 is selected, choose claims numbered 03, 10, 17, 24, 31 and 38).

Comment on each of the methods including an explanation of whether it would yield a random sample or not.

Solution

Method 1. This would be a sensible policy for the insurance company to adopt but it would not be a random sample. The smaller claims have no chance of being chosen.

Method 2. This is exactly the method we have used in Mandela Close and would yield a random sample.

Method 3. This would not give all claims an equal chance of being chosen. For example, the claim numbered 01 would be included in the sample if 01, 43 or 85 were selected. However, the claim numbered 30 would only be included if 30 or 72 were selected.

Method 4. This would yield a random sample. All claims would have two numbers associated with them and so have an equal chance of being chosen. In the example on method 3 the claim numbered 01 is now only chosen if 01 or 43 were selected.

Method 5. Each claim would have an equal chance of being selected but this would not be a random sample as not all combinations of six claims could be chosen.

Method 4 is unnecessarily complicated but has the advantage that less two-digit random numbers are rejected as too high than in method 2. It might be useful if there were, say, 1050 items in the population numbered 0000 to 1049. Four-digit random numbers would be needed but the great majority would be out of the required range.

EXERCISE 2A

1 On a particular day there are 2125 books on the shelves in the fiction section of a library. Describe how random numbers could be used to select a random sample of size 20 (without replacement) from the 2125 books. [A]

2 Describe how random numbers could be used to select a simple random sample of size 6 from the 712 employees of a large city centre store. [A]

3 A gardener grew 28 tomato plants. Describe how you would use random numbers to take a simple random sample of size 8 from the population. [A]

4 The ages in years of the students in a statistics class are given below.

19	20	23	21	21	20	20	19	19	20	19	24
20	19	20	22	21	25	20	33	19	19	19	20
24	36	27	33	26	38	43	24	41	30	27	49

Explaining fully the procedure you use take:
  (a) an unrestricted random sample (i.e. allow the same student to be chosen more than once) of size 6 from the population,
  (b) a simple random sample (i.e. do not allow the same student to be chosen more than once) of size 6 from the population. [A]

5 Describe how a simple random sample of 20 rods could be taken from a population of 500 rods. [A]

6 In order to estimate the mean number of books borrowed by members of a public library, the librarian decides to record the number of books borrowed by a sample of 40 members. She chooses the first member of the sample by selecting a random integer, r, between 1 and 5 inclusive. She then includes in her sample the rth member to leave the library one morning and every 5th member to leave after that until her sample of 40 is complete. Thus if $r = 3$ she chooses the 3rd, 8th, 13th … 198th members leaving the library as her sample.

    (a) Does the sample constitute a random sample of the first 200 people leaving the library? Give a reason.

A list of the names of the 8950 members of the library is available.

    (b) Describe how random sampling numbers could be used to select a random sample (without replacement) of 40 of these names.   [A]

7 In a particular parliamentary constituency there are 64 000 names on the electoral register. Of the electors, 32 000 live in property rented from the local authority, 21 000 live in owner-occupied property and 11 000 live in other types of property.

A total of 64 electors are selected at random from those living in property rented from the local authority, 42 electors are selected at random from those living in owner-occupied property and 22 electors are selected at random from those living in other types of property. State, giving a reason, whether a random sample of electors has been selected.

2.3 Stratified sampling

When sampling for opinion polls or social surveys there is usually a substantial amount of information available about the population which is being sampled. This information may come from the census, government publications or local authority records. It may be used to ensure that the sample is representative of the population. For example if we know that 34% of the population is aged over 60 we can ensure that 34% of the sample is aged over 60. This is only useful if the over-60s as a group have different opinions from the rest of the population. If there is no difference from the rest of the population then the extra work needed to take a stratified sample is done to no purpose.

It is generally understood that stratified sampling implies that a random sample is taken from each of the strata of the population. To emphasise this point this type of sampling is sometimes referred to as stratified random sampling.

This is **not** a random sample since all subsets of the population cannot be chosen. In a random sample it would be possible (but unlikely) for the whole sample to come from the same strata.

There may be circumstances where it would be desirable to 'over-represent' strata. For example to ensure that more than 34% of the sample is aged over 60. This is beyond the scope of this book.

A **stratified sample** requires prior knowledge to be used to divide the population into strata. Random samples are then taken from each of these strata, usually in proportion to the size of each of the strata.

 If relevant prior information about the population is available, a stratified sample using this information is preferable to a random sample.

Worked example 2.4

A gardener grew 28 tomato plants of which 21 were of strain *A* and seven of strain *B*. The gardener wishes to know the weight of tomatoes produced by the plants and decides to estimate this by weighing the yield of eight plants. Describe how random numbers could be used to select a stratified sample of size 8 from the 28 plants.

Solution

Of 28 plants, 21 are of strain *A*. So the sample should contain a proportion of $\dfrac{21}{28} = 0.75$ plants of strain *A*.

That is $0.75 \times 8 = 6$ plants of strain *A*.

Number the 21 plants of strain *A* from 00 to 20.

Select two-digit random numbers.

Ignore repeats and numbers greater than 20.

Continue until six two-digit numbers have been obtained. Select the corresponding plants.

To complete the sample we require two plants of strain *B*.

Number the seven plants of strain *B* from 0 to 6.

Select two-digit random numbers, ignoring repeats and numbers greater than six. When two have been obtained select the corresponding plants.

EXERCISE 2B

1 Explain briefly what you understand by a **stratified sample**. Give one advantage and one disadvantage of using stratified sampling compared to simple random sampling.

2 A German class consists of 12 male students and 18 female students. Describe how you would use random numbers to select a stratified sample of size 10 from this class.

3 A survey is to be carried out into the attitudes of voters in a parliamentary constituency to their member of parliament.

(a) Discuss the advantages and disadvantages of using a stratified sample if there is known to be:
  (i) a difference between the attitudes of men and women,
  (ii) no difference between the attitudes of men and women.

(b) Suggest three factors other than gender which might be used to stratify the population.

2.4 Quota sampling

If a survey of a large population, such as all adults living in the United Kingdom, is undertaken, it is usually impossible to achieve a random sample. There are difficulties in identifying the population, locating the individuals chosen and having located them persuading them to answer the questions. A **quota** sample is a stratified sample where no attempt is made to select the sample at random. The interviewers are given a quota of people to locate from given strata but are left to choose for themselves the particular individuals to include in the sample. For example an interviewer might be asked to interview eight male manual workers living in a suburban area.

Quota sampling is not as good as stratified random sampling because the interviewer will inevitably interview people who are easily accessible and willing to be interviewed. These people as a group may have different views from the population as a whole. However, quota sampling has been found to give useful results when the stratification is skilfully carried out.

> A **quota sample** is a stratified sample where the samples from each of the strata are chosen for convenience and there is no attempt at random sampling.

Worked example 2.5

In a large city 350 000 of the residents draw a state pension. These include 210 000 females of whom 80 000 are aged over 80. Of the males 20 000 are aged over 80.

A sample of 350 residents who draw a state pension is to be asked their views on the services for the elderly provided by the city.

(a) Give two advantages and one disadvantage of using a quota sample compared to a random sample.

(b) How many residents from each of the strata should be included in the sample and how should they be chosen?

Solution

(a) A quota sample is easier to carry out than a random sample (which would probably be impossible in practice).

Each of the strata would be fairly represented in the sample. This is desirable if different strata have different views.

Interviewers will choose people easily accessible to themselves which may introduce bias. A random sample would avoid this problem.

(b) The sample consists of 0.1% of the population and so a quota sample would consist of 0.1% of each of the strata.

i.e 80 females over 80

130 females not over 80

20 males over 80

120 males not over 80.

The samples from each of the strata should be chosen in any way convenient to the interviewer.

> If the numbers were not so convenient it might be impossible to take exactly the same proportion of each of the strata in the sample. It is possible to allow for different proportions from each of the strata in analysing the data. This is beyond the scope of this book.

EXERCISE 2C

1 Explain what you understand by a quota sample. Under what circumstances would you recommend the use of quota sampling?

2 A school has 1100 pupils of whom 320 are in the sixth form. A total of 180 of the sixth formers are girls as are 380 of the pupils who are not in the sixth form. A quota sample of size 55 is to be selected and asked to complete a questionnaire. Give detailed instructions to the school secretary who has no knowledge of sampling methods but has to decide which pupils are to be asked to complete the questionnaire.

3 A manufacturing company operates 15 factories, each with different numbers of employees. It has a total of 13 800 employees. The board of directors, concerned by a large turnover, decides to survey 100 employees to seek their opinions on working conditions. The following suggestions were made as to how the sample could be chosen.

Suggestion A. The employees are numbered 00 000 to 13 799. One hundred different five-digit random numbers between 00 000 and 13 799 are taken from random number tables and the corresponding employees chosen.

Suggestion B. The sample is made up of employees from all factories. The employees are selected at random from each factory, the number from each factory being proportional to the number of employees at the factory.

Suggestion C. The sample is made up of employees from all factories. The employees are selected by a convenient method, the number from each factory being proportional to the number of employees at the factory.

(a) (i) Which of the suggestions would produce a **quota** sample? [A]

    (ii) Name the type of sampling described in each of the other **two** suggestions. [A]

(b) For each of the **three** suggestions state whether or not all employees have an equal chance of being included in the sample.

(c) Give **one** reason for using:

    (i) **Suggestion A** in preference to **Suggestion C**,

    (ii) **Suggestion C** in preference to **Suggestion A**.

(d) Explain why **Suggestion B** might be preferred to **Suggestion A**.

2.5 Cluster sampling

Cluster sampling is used where it would be impractical to take a random sample but it is desirable to keep a random element in the sampling method. For example although it would be theoretically possible to interview a random sample of all junior school teachers in the United Kingdom it would involve an excessive amount of travelling and expense. Instead you could select a random sample of junior schools in the United Kingdom and then interview either all the teachers in these schools or a random sample of the teachers in the selected schools.

The amount of travelling would be greatly reduced. The disadvantage is that the teachers in a particular school are unlikely to hold views which are as varied as those of the population of all junior school teachers.

> Teachers in a particular school are likely to have views which are more **homogeneous** than the population of all teachers. The views of all teachers will be more **heterogeneous**.

There can be more than one step to cluster sampling. For example you could start by selecting a random sample of counties and then a random sample of junior schools within the selected counties. This would further cut down the travelling as the selected schools would now be clustered together rather than spread throughout the country.

> This is called **multi-stage** cluster sampling.

A **cluster sample** uses randomly chosen clusters of the population. It is not a random sample but contains an element of randomness.

Worked example 2.6

A health service union wishes to distribute a questionnaire to a sample of nurses working in hospitals. A list of all the hospitals in the United Kingdom is obtained and three are selected at random. Twenty nurses are chosen at random from each of the selected hospitals and asked to complete the questionnaire.

2

(a) What is the name given to this type of sampling? [A]

(b) Give **one** advantage and **one** disadvantage of this type of sampling. [A]

(c) Are all nurses working in hospitals in the United Kingdom equally likely to be included in the sample? Explain your answer. [A]

Solution

(a) Cluster sampling.

(b) Advantage – easier to administer as chosen nurses will be clustered in three hospitals instead of spread throughout the country.

Disadvantage – nurses in the same hospital likely to have less varied views than the whole population of nurses.

(c) No, nurses in small hospitals are more likely to be chosen than nurses in large hospitals.

The **hospitals** have an equal chance of being chosen.

EXERCISE 2D

1 Describe, with the aid of an example, the meaning of cluster sampling. Give one advantage and one disadvantage of cluster sampling compared to random sampling.

2 Explain, briefly, what you understand by cluster sampling. Under what circumstances would you advise its use?

Give one advantage and one disadvantage of cluster sampling compared to quota sampling.

3 The membership secretary of a football supporters club, which has 120 branches, wishes to contact a representative sample of members. She has a complete list of members classified by branches. She selects four branches at random and then ten members at random from each of the chosen branches.

(a) What name is given to this method of sampling?

(b) Would all members be equally likely to be included in the sample? Explain your answer.

(c) Under what (unlikely) circumstances would all members be equally likely to be included in the sample?

2.6 Systematic sampling

Sometimes items are selected on a regular pattern. For example every 100th vehicle from a production line could be tested or every 200th name on the electoral register could be chosen. In the case of the vehicles this ensures that the sample taken includes vehicles from throughout the day's production. It is not a random sample as, for example, it would be impossible for two consecutive vehicles to be both included in the sample.

Generally speaking systematic samples are perfectly satisfactory unless the pattern of the sampling follows a pattern in the population. For example if the first vehicle tested is always the 100th one produced each day it may be that the first few vehicles produced tend to be less satisfactory than the later ones and this would not be detected. Alternatively if it is known that every 100th vehicle is to be tested then extra care may be taken with these vehicles and so they are not typical of the whole population.

In a **systematic sample** members of the population are chosen at regular intervals.

Quota, cluster and systematic samples are used to overcome the practical problems associated with random sampling but a random sample is preferable.

Worked example 2.7

The 985 pupils of Aberdashers Comprehensive School are listed in alphabetical order and numbered, consecutively, from 000 to 984. A sample of 20 pupils is required to take part in a survey on eating habits.

A number between 0 and 34 is selected at random. The corresponding pupil and every 50th pupil thereafter is included in the sample (i.e. if the first number selected was 9, the pupils numbered 009, 059, 109 ... 909 and 959 would be selected).

(a) What is the name given to this method of sampling?

(b) Would all pupils be equally likely to be included in this sample? Explain your answer.

An alternative method of sampling is suggested. A number between 0 and 49 is selected at random. The corresponding pupil and every 50th pupil thereafter is included in the sample.

(c) **(i)** Would all pupils be equally likely to be included in this sample? Explain your answer.

(ii) Give a reason why neither of the two methods of sampling considered above would yield a random sample.

(iii) Give a reason, other than anything mentioned in your answers to **(c)(i)** or **(c)(ii)**, why the second method of sampling might be unsatisfactory in the given circumstances.

Solution

(a) Systematic sampling.

(b) No, pupils numbered 035 to 049 could not be included.

(c) **(i)** Yes, all pupils have a probability of $\frac{1}{50}$ of being included.

Pupils numbered 000 to 034 have a probability of $\frac{1}{35}$ of being chosen.

2

(ii) In neither case would it be possible for all sets of 50 pupils to form the sample. For example a brother and sister next to each other in the alphabetical list could never both be included in the same sample.

(iii) If the number selected at random was between 35 and 49 there would only be 19 pupils in the sample.

> Also in the first method of sampling all pupils do not have the same probability of being chosen.

EXERCISE 2E

1 Describe with the aid of a simple example what you understand by systematic sampling.

2 An alphabetical list of the 2700 employees of a distribution company is available. Describe how a systematic sample of size 50 could be selected.

3 In order to estimate the mean number of books borrowed by library users a librarian decides to record the number of books borrowed by a sample of 40 members using the library.
He selects a random integer, *r*, between 1 and 5. He includes in his sample the *r*th person to leave the library one morning and every 5th person after that until his sample of 40 is complete.

(a) What name is given to this method of sampling?

(b) Is each of the first 200 people leaving the library that morning equally likely to be included in the sample? Explain your answer.

(c) Explain why the sample does not constitute a random sample of the first 200 people leaving the library that morning.

(d) Comment on whether the sample will provide a useful estimate of the mean number of books borrowed by library users.

Worked example 2.8

On a particular day there are 2125 books on the shelves in the **fiction** section of a library.

(a) Describe how random numbers could be used to select a random sample of size 20 (without replacement) from the 2125 books. [A]

The number of times each book in the sample has been borrowed in the last year is found by counting the appropriate date stamps inside the front cover. The sample mean is used as an estimate of the mean annual number of times borrowed for all books belonging to the **fiction** section of the library.

(b) How is the estimate likely to be affected by the fact that:

(i) some books will not be on the shelves due to being in the possession of borrowers,

(ii) new books will have been on the shelves for less than a year? [A]

(c) Give a reason, other than those mentioned in **(b)**, why the sample will not be representative of **all** the books in the library. [A]

(d) Discuss, briefly, how the problems identified in **(b)** and **(c)** of estimating the mean number of times per year books in the library are borrowed could be overcome.

Solution

(a) Number books 0000 to 2124 either physically or by their position on the shelves.

Select four-digit random numbers.

Ignore repeats and > 2124.

Continue until 20 numbers obtained.

Select corresponding books.

(b) (i) The books in possession of borrowers are likely to be the more popular books and so borrowed more frequently on average. Estimate from books on shelves likely to be too low.

(ii) New books on shelves for less than a year will tend to lead to an underestimate of the mean number of times borrowed.

(c) Non-fiction books not included.

(d) Library may have computerised records which make sampling unnecessary. If not it will at least have a catalogue of all books and a random sample could be selected from this. Books in the sample which are out on loan could be checked when they are returned. The data for new books could be scaled up according to how long they had been on the shelves.

Worked example 2.9 _____

As part of an investigation of sampling methods, 150 plastic rods of varying lengths are placed on a desk by a teacher. A student is invited to view this population of rods. She is asked to estimate the mean length of the rods by choosing a sample of three which she thinks will have a mean length similar to that of the population (i.e. there is no attempt to select a random sample). The three selected rods are then measured and their mean length calculated. This process is repeated by each of the students in the class.

The following eight **mean** lengths, in centimetres, were obtained.

4.7 5.3 4.4 6.2 5.2 4.9 3.9 4.8

The mean length of the population of rods is 3.4 cm.

(a) Comment on this method of estimating the mean length of the population. [A]

2

(b) Use the eight mean lengths to estimate the mean and the standard deviation of the sample means. [A]

The teacher now asks a student to obtain a random sample of size three from the rod population.

(c) Describe how the student could use random number tables to obtain such a sample. [A]

Each student then independently obtains a random sample of size three from the population. The rods are measured and their mean lengths calculated as before. The eight **mean** lengths of the random samples are as follows:

4.6   1.9   7.2   1.5   2.2   2.8   3.7   1.4

(d) Use the eight mean lengths to estimate the mean and the standard deviation of the means of random samples of size three. [A]

(e) Compare the means of the non-random samples with those of the random samples. Your comments should include reference to the population mean. [A]

(f) Estimate the standard deviation of the population of rod lengths. [A]

(g) A sample of size 16 is taken from the population and the mean length is found to be 3.8 cm. It is claimed to be a random sample. State, giving a reason, whether or not this claim is reasonable.

Solution

(a) All eight sample means are above the population mean suggesting that the method is biased.

(b) 4.925, 0.680.

(c) Number rods 000 to 149.
Select three-digit random numbers.
Ignore 150 and over and repeats.
Continue until three numbers obtained.
Select corresponding rods.

(d) 3.16, 1.97

(e)
Random sample means	Non-random sample means
Some above population mean some below (unbiased)	All above population mean (biased)
Mean of means close to population mean.	Mean of means not close to population mean.
More variable than non-random.	Less variable than random

(f) For population standard deviation σ, the standard deviation of the mean of three randomly selected observations is $\dfrac{\sigma}{\sqrt{3}}$. Hence σ is estimated by $\sqrt{3} \times 1.97 = 3.41$.

> This requires knowledge from other areas of statistics. There will always be some questions on an examination paper which are not restricted to a single topic.

(g) Estimated standard deviation of mean of random sample of

size 16 is $\dfrac{3.41}{\sqrt{16}} = 0.85$.

Hence 3.8 cm is about 0.5 standard deviations above the mean. The claim is plausible.

> The population mean is given as 3.4 cm.

EXERCISE 2F

1 The ages, in years, of the students in a statistics class are given below. The letters 'pt' following an age indicate that the student was attending the course on a part-time basis. The other ages are for full-time students.

19	20	23	21	21	20	20	19	19	20	19	24
20	19	20	22	21	25	20	33	19	19	19	20
49pt	24pt	36pt	27pt	33pt	26pt	38pt	43pt	24pt	41pt	30pt	27pt

(a) Describe how random numbers could be used to take a stratified sample of size 12 from this class.

(b) The mean age of a stratified sample of size 12 is $\bar{x}$ and the mean age of a random sample of size 12 is $\bar{y}$. Which of $\bar{x}$ and $\bar{y}$ would you choose as an estimate of the class mean age? Explain your answer.

2 A fast-food shop employs a student to conduct a survey into the eating habits of its customers. The student uses random number tables to choose a number, r, between 1 and 8. On a particular day she arrives when the shop opens and asks the rth customer entering the shop to complete a questionnaire. She then asks every 8th customer who enters the shop to complete the questionnaire until she has asked a total of 50 customers.

(a) Are all the first 400 customers entering the shop that day equally likely to be asked to complete the questionnaire?

(b) Are the chosen customers a random sample of the first 400 customers entering the shop that day? Give a reason for your answer.

(c) Discuss briefly whether the views expressed by these customers are likely to be representative of the views of all customers.

3 In a particular parliamentary constituency there are 64 000 names on the electoral register. Of the electors, 32 000 live in property rented from the local authority, 21 000 live in owner-occupied property and 11 000 live in other types of property.

The following methods are suggested for choosing a sample of electors in order to carry out an opinion survey.

A Use a random process to select 128 names from the electoral register.

B   Use a random process to select one of the first 500 names on the electoral register. Using this as a starting point select every 500th name.

C   Select 64 names at random from the electors living in property rented from the local authority, 42 names at random from the electors living in owner-occupied property and 22 names at random from those living in other types of property.

(a) For **each** of the methods **A**, **B** and **C**:

    (i)   name the type of sampling method,

    (ii) state whether all the names on the electoral register are equally likely to be included in the sample.   [A]

(b) State, giving a reason, whether method **C** will produce a random sample of the electors.   [A]

(c) State, briefly, the difference between a sample obtained by method **C** and a quota sample. What is the advantage of a quota sample compared to a sample obtained by method **C**?   [A]

(d) Compare the usefulness of sampling methods **A** and **C** if the questions to be asked concerned local authority housing policy.   [A]

(e) How would your answer to **(d)** be changed (if at all) if the questions to be asked concerned attitudes to the monarchy? Explain your answer.   [A]

4   There are 28 houses in Mandela Road, 14 on each side. The houses on one side of the road have even numbers and those on the other side of the road have odd numbers.

A total of 63 residents of Mandela Road are on the electoral register.

A market researcher wishes to interview seven of these residents. He decides to choose a sample of seven houses using the following procedure:

Step 1   Toss a coin and choose the side of the road with odd numbered houses if it falls heads and the side of the road with even numbered houses if it falls tails.

Step 2   Toss the coin again and select the lowest numbered house on the chosen side of the road if it falls heads and the second lowest if it falls tails.

Step 3   Select alternate houses on the chosen side of the road starting from the house chosen in step 2. (For example, a tail followed by another tail would result in him selecting the houses numbered 4, 8, 12, 16, 20, 24 and 28.)

(a) (i) Would all houses in Mandela Road be equally likely to be included in the sample? Explain your answer. [A]

   (ii) Would the sample be random? Give a reason for your answer. [A]

The market researcher knocks at each selected house and asks the person who opens the door to answer a questionnaire. Assume that each person who opens the door is on the electoral register and is willing to answer the questionnaire.

(b) Give **two** reasons why the people who answer the questionnaire are not a random sample from the 63 residents on the electoral register. [A]

(c) Describe how random numbers could be used to select a random sample of size seven (without replacement) from the 63 residents on the electoral register. [A]

5 A small trade union has 23 000 members divided (unequally) into 12 branches. The executive committee decides to survey 200 members to seek their opinions of the services provided by the union. The following four suggestions were made as to how the sample should be chosen.

Suggestion A   Select 5 of the 12 branches at random. The sample consists of 40 members selected at random from each of the 5 chosen branches.

Suggestion B   The sample is made up of members from all branches. The members are selected at random from each branch, the number from each branch being proportional to the size of the branch membership.

Suggestion C   As Suggestion B except that, instead of members being selected at random, each branch secretary is asked to select a sample in the most convenient way.

Suggestion D   The members are numbered 00 000 to 22 999. Two hundred different 5-digit random numbers between 00 000 and 22 999 are taken from random number tables and the corresponding members chosen.

(a) State which of the suggestions would yield a quota sample. [A]

(b) For each of the other three suggestions state the type of sample described. [A]

(c) For each of the four suggestions state whether or not each member of the union would have an equal chance of being selected. [A]

2

(d) State, with a reason, which of the suggestions is best from

  (i) a statistical point of view,

  (ii) a practical point of view.

6 On a train from Manchester to London, 70 passengers have reserved first class seats and 180 passengers have reserved standard class seats. A list of these passengers is available.

(a) Describe how random numbers could be used to select a stratified sample of size 50 from the list.

(b) A survey of passengers on the train is to be undertaken. It is decided that the sample in **(a)** would not be useful as some people who reserve seats do not travel, whilst others travel without reserving seats. On this train, 216 people travelled in standard class accommodation. This accommodation consisted of three carriages. In each carriage there were 72 seats numbered from 1 to 72. All seats were occupied. As part of the survey, a sample of 45 passengers travelling in standard class accommodation is required. The following suggestions are made:

Suggestion A Select 2 digit random numbers, ignoring 00, numbers above 72 and any repeats. If the first number is, for example, 56, choose the 3 seats (one in each carriage) numbered 56. The passengers sitting in these seats are included in the sample. Continue with the next 2 digit random number until a sample of 45 passengers is obtained.

Suggestion B Number the 216 seats from 000 to 215. Select 3 digit random numbers, ignoring repeats and numbers greater than 215. The passengers sitting in the selected seats are included in the sample. Continue until a sample of 45 passengers is obtained.

  (i) Explain why one of the two suggestions leads to a random sample and the other does not.

  (ii) For the suggestion which does not lead to a random sample, state whether all of the 216 passengers travelling in standard class accommodation are equally likely to be included in the sample. Explain your answer.

7 A local education authority is responsible for 85 schools and employs 1800 teachers of whom 1100 are female and 700 are male. It wishes to carry out a survey of teachers' attitudes to maternity leave and decides to ask 90 teachers to agree to being interviewed. Two alternative methods of selecting the teachers are proposed.

Method 1 Select 55 teachers at random from the 1100 female teachers and 35 teachers at random from the 700 male teachers.

Method 2 Select ten schools at random from the 85 schools and 9 teachers at random from each of the ten schools.

(a) For each suggestion:
  (i) name the type of sampling described,
  (ii) state, giving a reason, whether or not every teacher would have an equal chance of being included in the sample.

(b) For Method 1:
  (i) give a reason why the sample selected will not be a random sample,
  (ii) bearing in mind the topic of the survey, give one advantage of this method of sampling,
  (iii) state the circumstances for which it would be unnecessarily complicated.

(c) Give one advantage and one disadvantage of the method of sampling described in Method 2.

Key point summary

1 A **random sample** of size n is a sample selected in such a way that all possible samples of size n have an equal chance of being selected. *p26*

2 A random sample chosen without replacement is called a **simple random sample**. A random sample chosen with replacement is called an **unrestricted random sample**. *p26*

3 A **stratified sample** requires prior knowledge to be used to divide the population into strata. Random samples are then taken from each of the strata, usually in proportion to the size of each of the strata. *p31*

4 If relevant prior information about the population is available, a stratified sample using this information is preferable to a random sample. *p31*

5 A **quota sample** is a stratified sample where the samples from each of the strata are chosen for convenience and there is no attempt at random sampling. *p32*

6 A **cluster sample** uses randomly chosen clusters of the population. It is not a random sample but contains an element of randomness. *p34*

7 In a **systematic sample** members of the population are chosen at regular intervals. *p36*

8 Quota, cluster and systematic samples are used to overcome the practical problems associated with random sampling but a random sample is preferable. *p36*

2

Test yourself	What to review
1 A list of the names of the 60 members of a cookery club is available. Describe how random numbers could be used to take a random sample of size 10.	*Section 2.2*
2 How could a systematic sample of size 10 be taken from the list in question **1**? How would you ensure that all members of the club had an equal chance of being included in such a sample?	*Section 2.6*
3 A firm employs 80 men and 140 women. Describe how random numbers could be used to take a stratified sample of size 22 from the employees. What difficulty would arise if the sample size required was 20? How would this difficulty be dealt with?	*Section 2.3*
4 How would a quota sample of size 22 be taken from the employees of the firm in question **3**?	*Section 2.4*
5 The members of the sample chosen in question **3** are asked how long they take to travel to work on a normal weekday. If the average time taken by the male and female employees is the same would a random sample be preferable to the sample chosen? Explain your answer.	*Section 2.3*
6 A journalist wishes to interview members of a political party which has branches in all parts of the country. A complete list of members is available. Explain to the journalist the advantages of choosing a cluster sample.	*Section 2.5*

Test yourself ANSWERS

1 Number names 00 to 59; select two-digit random numbers; ignore repeats and > 59; when 10 have been obtained choose corresponding names.

2 Choose a number between 00 and 05. Starting at this number choose every sixth name. If the original number is chosen by a random process all members of the club will have an equal chance of being selected.

3 Take a random sample of eight of the 80 men (as described in question **1**) and a random sample of 14 of the 140 women. If the total sample was of size 20 the samples from each of the strata could not be exactly proportional to the size of the strata. This could be allowed for in the analysis.

4 Choose eight of the 80 men and 14 of the 140 women in any convenient way.

5 There would be no advantage in taking a stratified sample compared to a random sample. The only disadvantage would be that it is more complicated to take a stratified sample.

6 A random sample would be impractical because of the amount of travelling which would be involved. A cluster sample where a few branches are chosen at random and members of these branches interviewed would avoid bias, by retaining a random element, but involve far less travelling.

CHAPTER 3

Discrete probability distributions

Learning objectives

After studying this chapter, you should be able to:

- find the expected value, standard deviation and variance of discrete random variables
- find the expected value of a function of a random variable.
- find the mode and the median of a discrete random variable.

3.1 Introduction

In S1 you have already met a discrete probability distribution – the binomial distribution.

It is possible to list the values a **discrete** variable may take together with their associated probabilities.

3.2 Expectation of a discrete random variable

The expected value of a discrete random variable, X, is usually denoted $E(X)$.

> A random variable is a variable whose value is (within limits) determined by chance.

> The expected value of a discrete random variable X is
>
> $$E(X) = \Sigma x P(X = x).$$

The summation is over all possible values of X.

In the AQA Formulae Book $P(X = x)$ is denoted P_i.

For example if X has the probability distribution below

x	0	2	8	10
$P(X = x)$	0.4	0.3	0.2	0.1

$$E(X) = (0 \times 0.4) + (2 \times 0.3) + (8 \times 0.2) + (10 \times 0.1) = 3.2$$

$E(X)$ is the mean score we would expect to obtain if samples were repeatedly taken from this distribution.

Random variables are usually denoted by an upper case letter such as X. Particular values of a random variable are usually denoted by a lower case letter such as x.

The expectation of g(X), where g(X) is any function of X is given by

$$E[g(X)] = \Sigma g(x)P(X = x).$$

For example, for the distribution above, if $g(X) = X^2$

$$E(X^2) = (0^2 \times 0.4) + (2^2 \times 0.3) + (8^2 \times 0.2) + (10^2 \times 0.1) = 24$$

3

3.3 Mean and variance of a discrete random variable

The mean of a discrete random variable is defined to be E(X).

The mean of X is E(X).

This is the definition of the mean of a random variable and is not the same as the definition of the mean of a set of data. However the two means are closely related.

For a set of data which consists of the observations $x_1, x_2, \ldots, x_n$ occuring with frequency $f_1, f_2, \ldots, f_n$, respectively, the mean is $\Sigma f_i x_i / \Sigma f_i$. This may be written $\Sigma x_i (f_i / \Sigma f_i)$. The definition of the mean of a probability distribution is obtained by replacing the relative frequency of occurrence of x_i, $(f_i / \Sigma f_i)$, by the probability of x_i.

For a similar reason the variance of a discrete random variable is defined to be $E[\{X - E(X)\}^2]$.

The variance of a discrete random variable is defined

$$Var(X) = E[\{X - E(X)\}^2].$$

For the distribution above, we have calculated E(X), the mean, as 3.2.

x	0	2	8	10
$(x - 3.2)^2$	10.24	1.44	23.04	46.24
$P(X = x)$	0.4	0.3	0.2	0.1

The variance of X:

$$Var(X) = (10.24 \times 0.4) + (1.44 \times 0.3) + (23.04 \times 0.2) + (46.24 \times 0.1) = 13.76$$

It is straightforward to show algebraically that

$$E[\{X - E(X)\}^2] = E(X^2) - [E(X)]^2.$$

You will not be asked to prove this.

Or $\sigma^2 = E(X^2) - \mu^2$.

This is an easier way of calculating the variance. We have already calculated $E(X) = 3.2$ and $E(X^2) = 24$.
Thus, $Var(X) = 24 - 3.2^2 = 13.76$ as before.

The variance plays an important role in mathematical statistics but the standard deviation is a more natural measure of spread. This is found by simply taking the square root of the variance.

In this case standard deviation $= \sqrt{13.76} = 3.71$.

Worked example 3.1

In a 'pay and display' car park, motorists use an automatic machine to purchase tickets. The price, £X, of the ticket depends on the length of time the motorist intends to leave the car. The following probability distribution provides a suitable model for the random variable X.

x	$P(X = x)$
0.8	0.25
1.4	0.55
2.0	0.08
2.8	p

(a) Find the value of p.
(b) Calculate the mean and the standard deviation of X.
(c) A few motorists park but do not purchase a ticket. It is decided to modify the probability distribution to include these motorists by allocating a small probability to the outcome $X = 0$. Will the standard deviation of the modified probability distribution be greater, the same or smaller than that calculated in (b)?

Solution

(a) $p = 1 - 0.25 - 0.55 - 0.08 = 0.12$

The sum of all the probabilities must equal one.

(b) $E(X) = (0.8 \times 0.25) + (1.4 \times 0.55) + (2.0 \times 0.08)$
$\qquad + (2.8 \times 0.12)$
$\qquad = 1.466$
$E(X^2) = (0.8^2 \times 0.25) + (1.4^2 \times 0.55) + (2.0^2 \times 0.08)$
$\qquad + (2.8^2 \times 0.12)$
$\qquad = 2.4988$
variance of $X = 2.4988 - 1.466^2 = 0.3496$
standard deviation $= \sqrt{0.3496} = 0.591$

Don't round to 3 s.f. at this stage. You will need to keep all these figures in order to obtain the standard deviation correct to 3 s.f.

(c) If $X = 0$ is included in the probability distribution it will be more spread out and so the standard deviation will be increased.

Worked example 3.2

A newsagent sells phone cards valued at £1, £2, £4, £10 and £20. The value, in £, of a phone card sold may be regarded as a random variable, X, with the following probability distribution:

x	$P(X = x)$
1	0.20
2	0.40
4	0.22
10	0.11
20	0.07

(a) Find the mean and standard deviation of X.

(b) What is the probability that the value of the next phone card sold is less than £4?

The newsagent is considering whether to discontinue selling £1 and £2 cards. If she did this a proportion, p, of customers who presently buy £1 or £2 cards would then buy a £4 card. Other such customers would not buy a card. As a result the value, in £, of sales would be a random variable, Y, with the following probability distribution:

y	$P(Y = y)$
0	$0.6(1 - p)$
4	$0.22 + 0.6p$
10	0.11
20	0.07

(c) Find the mean of Y in terms of p.

A survey suggests that the value of p would be between 0.5 and 0.7.

(d) Using this information, advise the newsagent on the likely effect on takings if she decides to discontinue selling £1 and £2 cards. Also point out factors, other than receipts from phone cards, which she should consider before making a decision. [A]

Solution

(a) $E(X) = (1 \times 0.2) + (2 \times 0.4) + (4 \times 0.22) + (10 \times 0.11) + (20 \times 0.07)$
$= £4.38$

$E(X^2) = (1^2 \times 0.2) + (2^2 \times 0.4) + (4^2 \times 0.22) + (10^2 \times 0.11)$
$+ (20^2 \times 0.07) = 44.32$

> Don't forget to square X, **not** the probability.

variance $= 44.32 - 4.38^2 = 25.1356$
standard deviation $= \sqrt{25.1356} = £5.01$

(b) Probability less than £4 $= 0.20 + 0.40 = 0.60$

(c) $E(Y) = (0 \times 0.6(1 - p)) + 4(0.22 + 0.6p) + (10 \times 0.11) + (20 \times 0.07)$
$= £(3.38 + 2.4p)$

(d) The survey suggests that p will be 0.5 or higher. The expected takings will be £4.58 or higher. Thus there will be an increase in takings from phone cards. However there may be a loss of goodwill and some customers who used to buy £1 and £2 phone cards may no longer come into the shop. This could lead to a reduction in the sales of other items.

Worked example 3.3

The four directors of a company each have a parking space reserved for them at head office. Based on past observations, the number of these spaces occupied at 10.00 a.m. on a weekday morning may be modelled by a random variable, R, with the following probability distribution:

r	0	1	2	3	4
$P(R = r)$	0.25	0.30	0.15	0.10	0.20

(a) Calculate the mean and standard deviation of R.

(b) It is suggested that a binomial distribution may provide an adequate model for R. Assuming that this is correct:

    (i) use the mean you have calculated in **(a)** to estimate p, the probability that a parking space is occupied at 10.00 a.m. on a weekday morning,

    (ii) use your estimate of p to estimate the standard deviation of R.

(c) Do your calculations support the suggestion that a binomial distribution provides an adequate model for R?

(d) Give two reasons why, irrespective of your calculations, a binomial distribution may not be an appropriate model for R. [A]

Solution

(a) $E(R) = (0 \times 0.25) + (1 \times 0.30) + (2 \times 0.15) + (3 \times 0.10) + (4 \times 0.20)$
$= 1.7$
$E(R^2) = (0^2 \times 0.25) + (1^2 \times 0.30) + (2^2 \times 0.15) + (3^2 \times 0.10) + (4^2 \times 0.20)$
$= 5.0$

    variance of $R = 5 - 1.7^2 = 2.11$
standard deviation $= \sqrt{2.11} = 1.45$

(b) (i) Mean of binomial is np. In this case the mean $= 1.7$ and $n = 4$, hence the estimate of

$$p = \frac{1.7}{4} = 0.425.$$

    (ii) Standard deviation of binomial is $\sqrt{np(1-p)}$.

Hence estimated standard deviation of R is $\sqrt{4 \times 0.425(1 - 0.425)} = 0.989$.

(c) The standard deviation calculated from the probability distribution is 1.45. This is not close to 0.989, the estimate based on assuming a binomial distribution. Hence calculations suggest that binomial is not an adequate model.

(d) p may not be constant. Different directors may have different probabilities of using car parking space at 10.00 a.m.

Use of car parking space may not be independent. A meeting on the premises may mean that all directors will be using their parking spaces whereas a meeting off the premises may mean that none of the directors will be using their parking spaces.

3.4 Mode and median of a discrete random variable

> The mode of a discrete random variable is the observation with the largest probability.

r	0	1	2	3	4	5
$P(R = r)$	0.11	0.25	0.21	0.19	0.16	0.08

The mode of R is 1.

> If m is the median of R, then $P(R < m) \leqslant 0.5$ and $P(R \leqslant m) \geqslant 0.5$.

This definition could lead to two possible values for the median. This is very unlikely to occur in an examination question. If it does, take the mean of the two possible values as the median.

For the distribution above, the following table may be formed by accumulating the probabilities.

r	0	1	2	3	4	5
$P(R < r)$	0.00	0.11	0.36	0.57	0.76	0.92

The median of R is 3.

Worked example 3.4

A village inn offers bed and breakfast and has four bedrooms available for customers. The number of bedrooms occupied each night may be modelled by the random variable, X, with the following probability distribution:

x	$P(X = x)$
0	0.22
1	0.25
2	0.20
3	0.14
4	0.19

(a) Calculate:
   (i) the mean of X,
   (ii) the standard deviation of X,
   (iii) $E(X^3)$.

(b) Find the probability that the number of occupied bedrooms:
   (i) is less than 3,
   (ii) greater than or equal to the mean,
   (iii) greater than or equal to the mode,
   (iv) greater than or equal to the median.

(c) Find, approximately, the probability that on a random sample of 120 nights the mean number of bedrooms occupied will be less than 2.

Solution

(a) **(i)** $E(X) = (0 \times 0.2) + (1 \times 0.25) + (2 \times 0.20) + (3 \times 0.14) + (4 \times 0.19)$
       $= 1.83$
   (ii) $E(X^2) = (0^2 \times 0.22) + (1^2 \times 0.25) + (2^2 \times 0.20) + (3^2 \times 0.14) + (4^2 \times 0.19)$
       $= 5.35$

   Variance $= 5.35 - 1.83^2 = 2.0011$
   Standard deviation $= \sqrt{2.0011} = 1.41$
   (iii) $E(X^3) = (0^3 \times 0.22) + (1^3 \times 0.25) + (2^3 \times 0.20) + (3^3 \times 0.14) + (4^3 \times 0.19)$
       $= 17.79$

(b) **(i)** Probability less than three is the probability of 0, 1 or 2
       $= 0.22 + 0.25 + 0.20 = 0.67$.
   (ii) The probability greater than or equal to the mean (1.83) is the probability of 2, 3 or 4 $= 0.20 + 0.14 + 0.19 = 0.53$.
   (iii) The mode is 1 (highest probability).
       Probability greater than or equal to 1 is $0.25 + 0.20 + 0.14 + 0.19 = 0.78$.
   (iv) The median is 2 since $P(X < 2) = 0.22 + 0.25 = 0.47$ and
       $P(X \leqslant 2) = 0.22 + 0.25 + 0.20 = 0.67$.

       Probability greater than or equal to 2 $= 0.20 + 0.14 + 0.19 = 0.53$.

(c) The distribution of the mean of a large sample may be approximated by a normal distribution, even though the distribution of X is discrete. The mean of this distribution will be 1.83 and the standard deviation $\sqrt{2.0011/120}$.   See S1, section 5.11.

$z = (2 - 1.83)/\sqrt{2.0011/120} = 1.316$

Probability mean number of rooms occupied is less than 2 is 0.906.

EXERCISE 3A

1 A discrete random variable, X, has probability distribution defined by

x	0	1	2	3
$P(X = x)$	0.4	0.3	0.2	0.1

Find the mean, variance and standard deviation of X.

2 A discrete random variable, X, has probability distribution defined by

x	0	1	4	10
$P(X = x)$	0.2	0.5	0.2	0.1

   (a) Find the mean, variance and standard deviation of X.

   (b) Find $E(X^3)$.

3 A discrete random variable, X, has probability distribution defined by

x	0	1	2	3	4
$P(X = x)$	0.2	0.4	0.2	0.1	p

   (a) Find p.

   (b) Find the mean, variance and standard deviation of X.

   (c) Find $E(X^4)$.

4 Members of a public library may borrow up to four books at any one time. The number of books borrowed by a member on each visit to the library is a random variable, X, with the following probability distribution:

x	$P(X = x)$
0	0.24
1	0.12
2	0.20
3	0.28
4	0.16

   (a) Find the mean and the standard deviation of X.

   (b) Verify that $E[(X - E(X))^2] = E(X^2) - [E(X)]^2$.

5 Find:

   (a) the mode of the distributions in questions **1**, **2** and **3**.

   (b) the median of the distributions in questions **1**, **2**, **3** and **4**.

6 A regular customer at a small clothes shop observes that the number of customers, X, in the shop when she enters has the following probability distribution:

Number of customers, x	$P(X = x)$
0	0.15
1	0.34
2	0.27
3	0.14
4	0.10

   (a) Find the mean and standard deviation of X.

She also observes that the average waiting time, Y, before being served is as follows:

Number of customers, x	Average waiting time, y minutes
0	0
1	2
2	6
3	9
4	12

(b) Find her mean waiting time.

7 Prospective recruits to a large retailing organisation undergo a medical examination. As part of the examination, their heights are measured by a nurse, and recorded to the nearest 2 mm. The final digit of the recorded height may be modelled by a random variable, X, with the following probability distribution:

x	0	2	4	6	8
$P(X = x)$	0.2	0.2	0.2	0.2	0.2

(a) Find the mean and standard deviation of X.

(b) A new nurse recorded the heights to the nearest 5 mm. Construct an appropriate probability distribution for the final digit of the recorded height. [A]

8 A company produces blue carpet material. The length of material (in metres) required to meet each order is a discrete random variable, X, with the following probability distribution:

x	$P(X = x)$
50	0.50
60	0.08
70	0.04
80	0.05
90	0.08
100	0.25

(a) Find the mean and the standard deviation of X.

(b) What is the probability that in a day during which exactly two orders are placed the total length of material ordered is 120 m? [A]

9 Applicants for a sales job are tested on their knowledge of consumer protection legislation. The test consists of five multiple choice questions. The number of correct answers, X, follows the distribution below.

X	0	1	2	3	4	5
$P(X = x)$	0.60	0.04	0.07	0.10	0.09	0.10

(a) Find the mean and standard deviation of X.

A group of production staff, who had no knowledge of the subject, guessed all the answers. The probability of each answer being correct was 0.25. The random variable, Y, represents the distribution of the number of correct answers for this group.

(b) **(i)** Name the probability distribution which could provide a suitable model for Y.

    (ii) Determine the mean and standard deviation of Y.

(c) Compare and comment briefly on the results of your calculations in **(a)** and **(b)(ii)**.   [A]

10 A petrol station in a remote area installs a self-service machine which delivers petrol on the insertion of £1 coins. This enables petrol to be obtained when the owner is not available. Observation suggests that the distribution of the value of petrol bought from the machine by each customer is as follows:

Value of petrol, £x	Probability
1	0.18
2	0.12
3	0.22
4	0.18
5	0.12
6	0.08
7	0.06
8	0.04

(a) Find the mean and standard deviation of X.

(b) The owner considers having the machine adjusted so that at least five £1 coins would have to be inserted to obtain petrol. Assume that the distribution of sales to present customers after a change would be as follows:

Value of petrol, £y	Probability
0	$0.7(1 - p)$
5	$0.12 + 0.7p$
6	0.08
7	0.06
8	0.04

Find the mean of Y in terms of p and hence the range of values of p which would lead to an increase in mean sales.   [A]

11 Marian belongs to the Handchester Building Society. She frequently visits her local branch to pay instalments on her mortgage. The number of people queuing to be served when she enters the branch may be modelled by the random variable X with the following probability distribution:

x	$P(X = x)$
0	0.12
1	0.33
2	0.27
3	0.18
4	0.07
5	0.03

(a) Find the probability that when she enters the branch there are two or more people queuing to be served.

(b) Find:
   (i) the mean of X,
   (ii) $E(X^2)$,
   (iii) the standard deviation of X.

(c) Find, approximately, the probability that in 100 visits to the branch the mean number of people queuing to be served when she enters is two or more.

12 The amount charged, £X, for entry to a museum depends on the status of the visitor. The following table shows the charges together with the probability that a visitor will have a particular status.

Status	Charge, £X	$P(X = x)$
Child under 16	1.00	0.35
Student	1.50	0.21
Senior citizen	2.00	0.24
Adult	3.00	0.20

(a) For entrance charges paid by visitors to the museum calculate:
   (i) the mean,
   (ii) $E(X^2)$,
   (iii) the standard deviation.

(b) Find the probability that the charge for a randomly selected visitor will be greater than or equal to:
   (i) the mean,
   (ii) the mode.

(c) Children under 5 are admitted free and have been omitted from the probability distribution shown above. If they were included in the probability distribution, explain whether:

(i) the mean would increase, stay the same or decrease,

(ii) the standard deviation would increase, stay the same or decrease. [A]

Key point summary

I For a discrete random variable, X,   *p16*

$$E(X) = \Sigma x P(X = x)$$

where the summation is over all possible values of X.

2 For a discrete random variable X,   *p47*

$$E[g(X)] = \Sigma g(x)P(X = x)$$

where the summation is over all possible values of X.

3 The mean of X is $E(X)$ *p47*

4 The variance of X,   *p47,48*

$$\mathbf{VAR}(X) = \mathbf{E}[\{X - E(X)\}^2] = \mathbf{E}(X^2) - [\mathbf{E}(X)]^2$$

5 The mode of a discrete random variable is the observation with the largest probability.   *p51*

6 If m is the median of R, $P(R < m) \leqslant 0.5$ and $P(R \leqslant m) \geqslant 0.5$.   *p51*

Test yourself	What to review

1 A discrete random variable, X, has a probability distribution defined by   *Section 3.2*

X	0	1	4	10
$P(X = x)$	0.1	0.3	0.4	0.2

Find:

(a) $E(X)$, **(b)** $E(X^2)$, **(c)** $E(X^3)$, **(d)** probability $X > 4$.

2 A discrete random variable, R, has a probability distribution defined by   *Sections 3.3 and 3.4*

r	2	3	4	5	6
$P(R = r)$	0.33	0.27	0.26	0.08	0.06

Find the mean, median, mode and standard deviation of R.

Test yourself (*continued*)	**What to review**

3 A discrete random variable, R, has the following probability distribution: — *Sections 3.2 and 3.3*

r	5	10	15	20
$P(R = r)$	0.15	0.25	0.40	0.20

(a) Find:

   (i) $E(R)$,

   (ii) $E(R^2)$,

   (iii) $E\{[(R - E(R)]^2\}$,

   (iv) Verify that $E\{[(R - E(R)]^2\} = E(R^2) - [E(R)]^2$.

(b) What is the name given to $E\{[(R - E(R)]^2\}$?

---

4 A discrete random variable, X, has probability distribution defined by — *Worked examples 3.1 and 3.4*

x	0	2	4	6	8
$P(X = x)$	p	$4p$	0.2	0.1	0.1

(a) Find p

(b) Find the probability that:

   (i) $X > 7.2$,

   (ii) $X < 2$,

   (iii) $X \leqslant 2$,

   (iv) $X >$ the median,

   (v) $X \geqslant$ the mode.

Test yourself ANSWERS

1 (a) 3.9; **(b)** 26.7; **(c)** 225.9; **(d)** 0.2.

2 mean 3.27, median 3, mode 2, s.d. 1.17.

3 (a) (i) 13.25, **(ii)** 198.75, **(iii)** 23.1875, **(iv)** $198.75 - 13.25^2 = 23.1875$ (answers not rounded to 3 s.f.); **(b)** variance.

4 (a) 0.12; **(b) (i)** 0.1, **(ii)** 0.12, **(iii)** 0.6, **(iv)** 0.4, **(v)** 0.88.

CHAPTER 4

Poisson distribution

Learning objectives

After studying this chapter, you should be able to:
- recognise circumstances where a Poisson distribution will provide a suitable model
- use tables of the Poisson distribution.

4.1 Introduction

> The Poisson distribution arises when events occur
> independently at random at a constant average rate.

For example the number of cars passing a point, per minute, on a quiet stretch of motorway might be modelled by a Poisson distribution. The number of telephone calls arriving at a switchboard over a 5-minute interval might also be modelled by a Poisson distribution. As for the binomial distribution (which you met in S1), only discrete, whole number outcomes are possible (0, 1, 2, 3, 4 …). However, unlike the binomial distribution there is no upper limit to the possible number of outcomes.

> A constant average rate does *not* mean that the same number of cars pass the point in each minute.

Other examples where the Poisson distribution might provide a suitable model are:

- the number of faults in a metre of dressmaking material
- the number of accidents per month on a particular stretch of motorway
- the number of daisies in a square metre of lawn.

4.2 The Poisson distribution

The French mathematician Siméon Denis Poisson showed that if events occur, in a given interval, independently at random at a constant average rate λ (that is if they follow a Poisson distribution), the probability that exactly r events will occur in a particular interval is:

$$\frac{e^{-\lambda}\lambda^r}{r!}$$

> This distribution is sometimes denoted Po(λ).

> e is an irrational number. Its value is approximately 2.718. You will **not** be asked to use this formula to evaluate Poisson probabilities in the SS02 exam.

4.3 Tables of the Poisson distribution

It is often unnecessary to use the Poisson formula because tables of the cumulative Poisson distribution are available (Table 2 in the Appendix). As with the binomial tables these tabulate the probability of 'x or fewer' events occurring. An extract is shown below.

Cumulative Poisson Distribution Function

The tabulated value is $P(X \leqslant x)$, where x has a Poisson distribution with mean λ.

x \ λ	0.1	0.2	0.3	0.4	0.5	0.6	0.7	0.8	0.9	0.10	1.2	1.4	1.6	1.8	λ \ x
0	0.9048	0.8187	0.7408	0.6703	0.6065	0.5488	0.4966	0.4493	0.4066	0.3679	0.3012	0.2466	0.2019	0.1653	0
1	0.9953	0.9825	0.9631	0.9384	0.9098	0.8781	0.8442	0.8088	0.7725	0.7358	0.6626	0.5918	0.5249	0.4628	1
2	0.9998	0.9989	0.9964	0.9921	0.9856	0.9769	0.9659	0.9526	0.9371	0.9197	0.8795	0.8335	0.7834	0.7306	2
3	1.000	0.9999	0.9997	0.9992	0.9982	0.9966	0.9942	0.9909	0.9865	0.9810	0.9662	0.9463	0.9212	0.8913	3
4		1.000	1.000	0.9999	0.9998	0.9996	0.9992	0.9986	0.9977	0.9963	0.9923	0.9857	0.9763	0.9636	4
5				1.000	1.000	1.000	0.9999	0.9998	0.9997	0.9994	0.9985	0.9968	0.9940	0.9896	5
6							1.000	1.000	1.000	0.9999	0.9997	0.9994	0.9987	0.9974	6
7										1.000	1.000	0.9999	0.9997	0.9994	7
8												1.000	1.000	0.9999	8
9														1.000	9

For example for a Poisson distribution with mean 0.9 the probability of two or fewer events occurring is 0.9371.

Worked example 4.1

On average eight vehicles pass a point on a free-flowing motorway in a 10-second interval. Find the probability that in a particular 10-second interval the number of cars passing this point is:

(a) seven or fewer,

(b) 12 or more,

(c) fewer than nine,

(d) more than eight,

(e) exactly nine,

(f) between seven and 10, inclusive.

> We can use the Poisson distribution here since we know the vehicles arrive at a constant average rate and that the traffic is free-flowing so that vehicles arrive independently and at random during this time interval.

Solution

(a) $P(X \leqslant 7) = 0.4530$
$\qquad = 0.453$

(b) $P(X \geqslant 12) = 1 - P(X \leqslant 11)$
$\qquad = 1 - 0.8881$
$\qquad = 0.1119$
$\qquad = 0.112$

> Here we use the $\lambda = 8.0$ column of the Poisson distribution tables.

> 11 | 12 13

(c) $P(X < 9) = P(X \leq 8)$
$= 0.5925$

| 7 | 8 | 9 | 10 |

(d) $P(X > 8) = 1 - P(X \leq 8)$
$= 1 - 0.5925$
$= 0.4075$

| 7 | 8 | 9 | 10 |

(e) $P(X = 9) = P(X \leq 9) - P(X \leq 8)$
$= 0.7166 - 0.5925$
$= 0.1241$
$= 0.124$

| 7 | 8 | 9 | 10 | 11 |

(f) $P(7 \leq X \leq 10) = P(X \leq 10) - P(X \leq 6)$
$= 0.8159 - 0.3134$
$= 0.5025$

| 5 | 6 | 7 | 8 | 9 | 10 | 11 | 12 |

4

EXERCISE 4A

1 The number of telephone calls arriving at a switchboard follow a Poisson distribution with mean 6 per 10-minute interval. Find the probability that the number of calls arriving in a particular 10-minute interval is:

(a) eight or fewer,

(b) more than three,

(c) fewer than six,

(d) between four and seven inclusive,

(e) exactly seven.

2 The number of customers arriving at a supermarket checkout in a 10-minute interval may be modelled by a Poisson distribution with mean 4. Find the probability that the number of customers arriving in a specific 10-minute interval is:

(a) two or fewer,

(b) fewer than seven,

(c) exactly three,

(d) between three and seven inclusive,

(e) five or more.

3 People arrive at a ticket office independently, at random, at an average rate of 12 per half-hour. Find the probability that the number of people arriving in a particular half-hour interval is:

(a) exactly 10,

(b) fewer than eight,

(c) more than 16,

(d) between nine and 15 inclusive,

(e) 14 or fewer.

4 The number of births announced in the personal column of a local weekly newspaper may be modelled by a Poisson distribution with mean 9.5. Find the probability that the number of births announced in a particular week will be:

(a) fewer than five,

(b) between six and 12 inclusive,

(c) eight or more,

(d) exactly 11,

(e) more than 11.

5 The number of people joining a checkout queue at a supermarket may be modelled by a Poisson distribution with a mean of 1.8 per minute. Find the probability that in a particular minute the number of people joining the queue is:

(a) one or fewer,

(b) exactly three.

6 The number of letters of complaint received by a department store follows a Poisson distribution with a mean of 6.5 per day. Find the probability that on a particular day:

(a) 7 or fewer letters of complaint are received,

(b) exactly 7 letters of complaint are received.

7 The weekly number of ladders sold by a small DIY shop can be modelled by a Poisson distribution with a mean of 1.4. Find the probability that in a particular week the shop will sell:

(a) 2 or fewer ladders,

(b) exactly 4 ladders,

(c) 2 or more ladders.

Worked example 4.2

As part of a feasibility study into introducing tolls on a motorway it is estimated that the number of cars arriving at a toll-booth site could be modelled by a Poisson distribution with mean 3.4 per 10-second interval. It is recommended that k toll-booths be installed where the number of cars arriving is $\leqslant k$ in at least 85% of 10-second intervals. Find k.

Solution

This question is answered by reading down the column $\lambda = 3.4$ until we find the first probability greater than 0.85. In this case we find that the probability of five or fewer cars arriving in a 10-second interval is 0.8705. Hence the required value of k is 5.

EXERCISE 4B

1 The number of customers arriving at an office selling tickets for a festival, may be modelled by a Poisson distribution with mean 1.2 per 2-minute interval. Find the number of arrivals which will not be exceeded in at least 90% of 2-minute intervals.

2 A garage offering a quick change service for exhausts finds that the demand for exhausts to fit a Metro may be modelled by a Poisson distribution with mean 8 per day. Find the demand which will not be exceeded on at least:

(a) 95% of days,

(b) 99% of days,

(c) 99.8% of days.

3 A small newsagent finds that weekday demand for the *Independent* follows a Poisson distribution with mean 12. How many *Independents* should the newsagent stock if the demand is to be satisfied on at least:

(a) 90% of days,

(b) 99% of days?

4 Demand for an item in a warehouse may be modelled by a Poisson distribution with mean 14 per day. The warehouse can only be stocked at the beginning of each day. How many items should the stock be made up to in order to ensure that demand can be met on:

(a) 95% of days,

(b) 99% of days,

(c) 99.5% of days?

4.4 The sum of independent Poisson distributions

If cars, going north, pass a point on a motorway independently at random and cars, going south, on the same motorway pass the point independently at random then all cars on the motorway will pass the point independently at random. If the cars going north follow a Poisson distribution with mean 5 per minute and cars going south follow a Poisson distribution with mean 6 per minute then the total number of cars passing the point will follow a Poisson distribution with mean $5 + 6 = 11$ per minute. This is a particular example of the general result that:

> If $X_1, X_2, X_3, \ldots$ follow independent Poisson distributions with means $\lambda_1, \lambda_2, \lambda_3, \ldots$ respectively, then
> $X = X_1 + X_2 + X_3 + \ldots$ follows a Poisson distribution with mean $\lambda = \lambda_1 + \lambda_2 + \lambda_3 + \ldots$.

In the example above of cars passing a point on a motorway, X_1 and X_2 are from different Poisson distributions. However, the result also applies if they are from the same distribution. For example X_1 could be the number of cars, going north, passing the point in a one minute interval and X_2 the number of cars, going north, passing the point in the next one minute interval. Both would be from a Poisson distribution with mean 5. $X_1 + X_2$ would be the number of cars, going north, passing the point in a 2-minute interval and would follow a Poisson distribution with mean $5 + 5 = 10$.

Worked example 4.3

The sales of a particular make of video recorder at two shops which are members of the same chain follow independent Poisson distributions with means 3 per day at the first shop and 4.5 per day at the second shop. Find the probability that on a given day:

(a) the first shop sells more than five,

(b) the second shop sells more than five,

(c) the total sales by the two shops is:

    (i)   five or fewer,

    (ii) more than 10,

    (iii) between five and nine inclusive.

Solution

(a) $P(X_1 > 5) = 1 - P(X_1 \leqslant 5)$
$$= 1 - 0.9161$$
$$= 0.0839$$

(b) $P(X_2 > 5) = 1 - P(X_2 \leqslant 5)$
$$= 1 - 0.7029$$
$$= 0.297$$

(c) Total sales of the two shops is Poisson with mean $3 + 4.5 = 7.5$.

    (i) $P(X \leqslant 5) = 0.2414$
$$= 0.241$$

    (ii) $P(X > 10) = 1 - P(X \leqslant 10)$
$$= 1 - 0.8622$$
$$= 0.1378$$
$$= 0.138$$

    (iii) $P(5 \leqslant X \leqslant 9) = P(X \leqslant 9) - P(X \leqslant 4)$
$$= 0.7764 - 0.1321$$
$$= 0.6443$$
$$= 0.644$$

Worked example 4.4 _____

A 1-day course on statistics for teachers of A-level geography is first advertised 8 weeks before it is due to take place.

Throughout these 8 weeks, the number of places booked follows a Poisson distribution with mean 2 per week.

(a) Find the probability that, during the first week, two or fewer places are booked.

The organisers are hoping for at least 20 participants. They decide that, if at the end of the first 5 weeks less than 10 places have been booked, then they will cancel the guest speaker.

(b) Find the probability that the guest speaker will be cancelled.

(c) Find the probability of exactly nine places being booked during the first 5 weeks.

(d) Exactly nine places were booked during the first 5 weeks. Find the probability that sufficient places are booked in the remaining 3 weeks to give a total of 20 or more bookings during the 8 week period.   [A]

Solution

(a) $P(X \leqslant 2) = 0.6767$

$\qquad = 0.677$

(b) The number of bookings in 5 weeks will follow a Poisson distribution with mean $2 + 2 + 2 + 2 + 2 = 10$

$\qquad P(X \leqslant 9) = 0.4579$

$\qquad\qquad = 0.458 \rightarrow$ guest speaker cancelled

(c) $P(X = 9) = 0.4579 - 0.3328$

$\qquad = 0.1251$

$\qquad = 0.125$

(d) 20 or more in 8 weeks $\rightarrow$ 11 or more in last 3 weeks.
Number of bookings in 3 weeks will follow a Poisson distribution with mean $3 \times 2 = 6$.

$\qquad P(X \geqslant 11) = 1 - P(X \leqslant 10)$

$\qquad\qquad = 1 - 0.9574$

$\qquad\qquad = 0.0426$

EXERCISE 4C

1 The sales of cricket bats in a sports shop may be modelled by a Poisson distribution with mean 1.2 per day. Find the probability that:

   (a) two or more bats are sold on a particular day,

   (b) eight or more bats are sold in a 5-day period,

   (c) exactly seven bats are sold in a 5-day period,

   (d) between four and seven bats are sold in a 5-day period.

2 Calls arrive at a switchboard independently at random at an average rate of 1.4 per minute. Find the probability that:

   (a) more than two calls will arrive in a particular minute,

   (b) more than 10 calls will arrive in a 5-minute interval,

   (c) between five and nine calls, inclusive, will arrive in a 5-minute interval,

   (d) more than 20 calls will arrive in a 10-minute interval,

   (e) 10 or fewer calls will arrive in a 10-minute interval.

3 The flaws in cloth produced on a loom may be modelled by a Poisson distribution with mean 0.4 per metre. Find the probability that there will be:

   (a) two or fewer flaws in a metre of this cloth,

   (b) more than three flaws in a 5-metre length of this cloth,

   (c) between three and six flaws, inclusive, in a 10-metre length of this cloth,

   (d) fewer than 10 flaws in a 30-metre roll of this cloth.

4 Two types of parasite were found on fish in a pond. They were distributed independently, at random with a mean of 0.8 per fish for the first type and 2.0 per fish for the second type. Find the probability that a fish will have:

   (a) three or fewer parasites of the first type,

   (b) more than one parasite of the second type,

   (c) a total of three or fewer parasites,

   (d) a total of exactly three parasites,

   (e) a total of more than five parasites.

5 A garage has two branches. The sales of batteries may be modelled by a Poisson distribution with mean 2.4 per day at the first branch and by a Poisson distribution with mean 1.6 per day at the second branch. Find the probability that there will be:

(a) exactly three batteries sold at the first branch on a particular day,

(b) exactly three batteries sold at the second branch on a particular day,

(c) a total of five or more batteries sold at the two branches on a particular day,

(d) a total of between one and five, inclusive, batteries sold at the two branches on a particular day,

(e) 15 or fewer batteries sold at the first branch in a 5-day period,

(f) more than eight batteries sold at the second branch in a 5-day period.

4.5 Using the Poisson distribution as a model

The Poisson distribution occurs when events occur independently, at random, at a constant average rate. A common example is cars passing a point on a motorway. However, if the motorway was busy, cars would obstruct each other and so would not pass at random. Hence the Poisson distribution would not be a suitable model. Also if we observed over a 24-hour period the mean number of cars per minute would not be constant. In the middle of the night the mean would be less than in the middle of the day. Again the Poisson distribution would not provide a suitable model. It probably would provide a suitable model if we counted the number of cars per minute on a free-flowing motorway over a relatively short period of time.

Telephone calls arriving at a switchboard are also often modelled by a Poisson distribution. However, if there were a queueing system or the switchboard was frequently engaged the calls would not be arriving independently at random. Also over a 24-hour period the mean would change and so the Poisson distribution would not be a suitable model. The Poisson distribution will often provide an adequate model for people joining queues in a supermarket or at a train station. However, if a family of four are shopping together they will probably all join the queue at the same time and so events will not be independent. Again if the queue is long people may be deterred from entering the supermarket and so once again the events would not be independent.

Despite all these qualifications the Poisson distribution provides a useful model in many practical situations. As with any probability distribution, we can never prove that events follow a Poisson distribution exactly, but we can recognise circumstances where it is likely to provide an adequate model.

EXERCISE 4D

State whether or not the Poisson distribution is likely to provide a suitable model for the random variable X in the following examples. Give a reason where you believe the Poisson distribution would not provide a suitable model.

The random variable X represents the number of:

(a) lorries per minute passing a point on a quiet motorway over a short period of time,

(b) lorries per minute passing a point on a very busy motorway,

(c) cars per minute passing a point close to traffic lights on a city centre road,

(d) components which do not meet specification in a sample of 20 from a production line,

(e) dandelions in a square metre of lawn in a small garden,

(f) boxes of expensive chocolates sold per day at a small shop,

(g) passengers per minute arriving at a bus stop, over a short period of time,

(h) passengers per minute arriving at a bus stop over a 24-hour period,

(i) breakdowns of a power supply per year,

(j) accidents per year in a large factory,

(k) people injured per year in accidents at a large factory.

Worked example 4.5

Travellers arrive at a railway station, to catch a train, either alone or in family groups. On an August Saturday afternoon, the number, X, of travellers who arrive alone during a one-minute interval may be modelled by a Poisson distribution with mean 7.5.

(a) Find the probability of six or fewer passengers arriving alone during a particular minute.

The number, Y, of family groups who arrive during a one-minute interval may be modelled by a Poisson distribution with mean 2.0.

(b) Find the probability that three or more family groups arrive during a particular minute.

It is usual for one person to buy all the tickets for a family group. Thus the number of people, Z, wishing to buy tickets during a one-minute interval may be modelled by $X + Y$.

(c) Find the probability that more than 18 people wish to buy tickets during a particular minute.

If four booking clerks are available, they can usually sell tickets to up to 18 people during a minute.

(d) State, giving a reason in **each** case, whether:

    (i) more than four booking clerks should be available on an August Saturday afternoon,

    (ii) the Poisson distribution is likely to provide an adequate model for the total number of travellers (whether or not in family groups) arriving at the station during a 1-minute interval,

    (iii) the Poisson distribution is likely to provide an adequate model for the number of passengers, travelling alone, leaving the station, having got off a train, during a 1-minute interval.

(e) Give **one** reason why the model $Z = X + Y$, used in **(c)**, may not be exact.

Solution

(a) $P(X \le 6) = 0.3782$
$$= 0.378$$

(b) $P(X \ge 3) = 1 - P(\le 2)$
$$= 1 - 0.6767 = 0.3233$$
$$= 0.323$$

(c) $X + Y \rightarrow$ Poisson mean $7.5 + 2 = 9.5$
$$P(X > 18) = 1 - P(X \le 18)$$
$$= 1 - 0.9957 = 0.0043$$

Only two significant figures in the answer can be obtained from the tables. In these circumstances a two significant figure answer will be accepted in an examination.

(d) (i) No, the probability of more than 18 people wishing to buy tickets in a minute has been shown in **(c)** to be very small. Hence four booking clerks should be adequate.

(ii) No, the people arriving in family groups will not be arriving independently. Hence Poisson unlikely to be an adequate model.

(iii) No, the average rate will not be constant. Immediately after a train arrives there will be a high average rate, between train arrivals there will be a low average rate.

(e) Some people or groups may have bought tickets in advance/some family groups may buy tickets individually.

4.6 Variance of a Poisson distribution

The Poisson distribution has the interesting property that the variance is equal to the mean. That is, a Poisson distribution, with mean 9, will have a variance of 9. You will probably be more interested in the standard deviation which will be $\sqrt{9} = 3$.

> You will remember from SS01 that the variance is used by theoretical statisticians but is a poor measure of spread.

A Poisson distribution with mean λ has a variance of λ (and a standard deviation of $\sqrt{\lambda}$).

The number of items of post delivered to a particular address, daily, follows a Poisson distribution with mean 9. On 10 days the number of items delivered was

10　8　13　9　4　9　12　15　11　8

> mean $\bar{x} = 9.9$
> standard deviation $s = 3.07$
> variance $s^2 = 9.43$

This is a sample from a Poisson distribution with mean 9 and so the mean of the sample will almost certainly not be exactly 9 nor will the variance be exactly 9. However for a large sample we would expect the mean and the variance both to be very close to 9.

This result will be used in later modules but for now its only application is to provide you with an extra piece of information when deciding whether or not the Poisson distribution may provide an adequate model. If you expected the number of items of post delivered per day to follow a Poisson distribution, then the fact that the mean (9.9) and variance (9.43) of the sample were close together would support this. Suppose the number of letters delivered on a sample of ten days had been:

12　6　15　7　2　7　14　17　13　6

with a mean of 9.9 and a variance of 24.1. You would have to say that it was very unlikely that the Poisson distribution would provide an adequate model as the mean and variance are so far apart.

MIXED EXERCISE

1 A small shop stocks expensive boxes of chocolates whose sales may be modelled by a Poisson distribution with mean 1.8 per day. Find the probability that on a particular day the shop will sell:

   (a) no boxes,

   (b) three or more boxes of these chocolates.   [A]

2 The number of births announced in the personal column of a local weekly newspaper may be modelled by a Poisson distribution with mean 2.4.

   Find the probability that, in a particular week:

   (a) three or fewer births will be announced,

   (b) exactly four births will be announced.   [A]

3 The number of customers entering a certain branch of a bank on a Monday lunchtime may be modelled by a Poisson distribution with mean 2.4 per minute.

   Find the probability that, during a particular minute, four or more customers enter the branch.   [A]

4 A shop sells a particular make of video recorder.

   (a) Assuming that the weekly demand for the video recorder is a Poisson variable with mean 3, find the probability that the shop sells:

      (i) at least three in a week,

      (ii) at most seven in a week,

      (iii) more than 20 in a month (4 weeks).

   Stocks are replenished only at the beginning of each month.

   (b) Find the minimum number that should be in stock at the beginning of a month so that the shop can be at least 95% sure of being able to meet the demand during the month.

5 Incoming telephone calls to a school arrive at random times. The average rate will vary according to the day of the week. On Monday mornings in term time there is a constant average rate of four per hour. What is the probability of receiving:

   (a) six or more calls in a particular hour,

   (b) three or fewer calls in a particular period of two hours?

   During term time on Friday afternoons the average rate is also constant and it is observed that the probability of no calls being received during a particular hour is 0.202. What is the average rate of calls on Friday afternoons?   [A]

6 State, giving a reason, whether or not the Poisson distribution is likely to provide an adequate model for the following distributions.

   (a) A transport cafe is open 24 hours a day. The number of customers arriving in each 5-minute period of a particular day is counted.

   (b) Following a cup semi-final victory, a football club ticket office receives a large number of telephone enquiries about tickets for the final, resulting in the switchboard frequently being engaged. The number of calls received during each 5-minute period of the first morning after the victory is recorded.

   (c) A machine produces a very large number of components of which a small proportion are defective. At regular intervals samples of 150 components are taken and the number of defectives counted.

7 A car-hire firm finds that the daily demand for its cars follows a Poisson distribution with mean 3.6.

   (a) What is the probability that on a particular day the demand will be:

      (i) two or fewer,

      (ii) between three and seven (inclusive),

      (iii) zero?

   (b) What is the probability that 10 consecutive days will include two or more on which the demand is zero?

   (c) Suggest reasons why daily demand for car hire may not follow a Poisson distribution. [A]

8 The number of letters received by a household on a weekday follows a Poisson distribution with mean 2.8.

   (a) What is the probability that on a particular weekday the household receives three or more letters?

   (b) Explain briefly why a Poisson distribution is unlikely to provide an adequate model for the number of letters received on a weekday throughout the year. [A]

9 The number of telephone calls to a university admissions office is monitored.

   (a) During working hours in January the number of calls received follows a Poisson distribution with mean 1.8 per 15-minute interval. During a particular 15-minute interval:

      (i) what is the probability that two or fewer calls are received,

      (ii) what number of calls is exceeded with probability just greater than 0.01?

(b) On any particular working day the number of attempts to telephone the office is distributed at random at a constant average rate. Usually an adequate number of staff are available to answer the telephone. However, for a short period in August, immediately after the publication of A-level results, the number of calls increases and the telephones are frequently engaged.

State, giving a reason, whether the Poisson distribution is likely to provide an adequate model for each of the following distributions:

(i) the number of calls received in each minute during working hours of a day in June,

(ii) the number of calls received in each minute during working hours of a day immediately after the publication of A-level results,

(iii) the number of calls received on each working day throughout the year. [A]

10 Bronwen runs a post office in a large village. The number of registered letters posted at this office may be modelled by a Poisson distribution with mean 1.4 per day.

(a) Find the probability that at this post office:

(i) two or fewer registered letters are posted on a particular day,

(ii) a total of four or more registered letters are posted on two consecutive days.

(b) The village also contains a post office run by Gopal. Here the number of registered letters posted may be modelled by a Poisson distribution with mean 2.4 per day. Find the probability that, on a particular day, the number of registered letters posted at Gopal's post office is less than 4.

(c) The numbers of registered letters posted at the two post offices are independent. Find the probability that, on a particular day, the total number of registered letters posted at the two post offices is more than six.

(d) Give one reason why the Poisson distribution might not provide a suitable model for the number of registered letters posted daily at a post office. [A]

11 The number of cars, travelling from East to West, passing a point on a motorway, may be modelled by a Poisson distribution with a mean of 1.2 per 5-second interval.

(a) Find the probability that, during a particular 5-second interval, the number of cars, travelling from East to West, which pass the point is:

(i) zero,

(ii) exactly two.

The number of cars, travelling from West to East, passing the same point on the motorway, may be modelled by a Poisson distribution with a mean of 3.8 per 5-second interval.

(b) Find the probability that, during a particular 5-second interval, more than eight cars, travelling from West to East, pass the point.

(c) Find the probability that, during a particular 5-second interval, the total number of cars passing the point is less than eight. (You may assume that the number of cars travelling from East to West is independent of the number of cars travelling from West to East.)

(d) Explain why a Poisson distribution may not provide an adequate model for the total number of car passengers passing the point in a 5-second interval. [A]

Key point summary

1 The Poisson distribution is the distribution of events which occur independently, at random, at a constant average rate.   *p59*

2 For a constant average rate λ, the probability of r events occurring, $P(R = r) = e^{-\lambda}\dfrac{\lambda^r}{r!}$.   *p59*

3 Dependent on the value of λ, Poisson probabilities may be found from tables.   *p60*

4 If $X_1, X_2, X_3, \ldots$ follow independent Poisson distributions with means $\lambda_1, \lambda_2, \lambda_3, \ldots$, respectively, then $X = X_1 + X_2 + X_3 + \ldots$ follows a Poisson distribution with mean $\lambda = \lambda_1 + \lambda_2 + \lambda_3 + \ldots$.   *p63*

5 A Poisson distribution with mean λ has a variance of λ.   *p70*

Test yourself   **What to review**

1 A Poisson distribution has a mean of 1.2 events per minute. Find, from tables, the probability that in a particular minute: *Section 4.3*

  (a) three or fewer events occur,

  (b) exactly three events occur,

  (c) less than three events occur,

  (d) more than three events occur.

2 What is the largest number of events which could occur in a given minute for the Poisson distribution in question **1**?   *Section 4.1*

Test yourself (*continued*)	What to review

3 The number of calls received at a switchboard may be modelled by a Poisson distribution with mean 12 per hour. Find the probability that more than one call will be received in a particular 5-minute interval.

Section 4.3

4 State two conditions which must be fulfilled if the number of bicycles crossing a bridge per minute is to follow a Poisson distribution.

Sections 4.1 and 4.5

5 Explain why the Poisson distribution is unlikely to form an adequate model for the number of bicycles crossing a bridge

(a) over a 24-hour period,

(b) in the rush hour when a large number of bicycles are attempting to cross.

Sections 4.1 and 4.5

4

6 The number of newspapers sold by a newsagent in eight successive hours on a weekday was

84   92   22   12   13   9   8   104.

(a) Calculate the mean and variance of the data.

(b) Give a reason, based on your calculations, why it is unlikely that the Poisson distribution will provide an adequate model for the data.

(c) Give a reason, not based on your calculations, why it is unlikely that a Poisson distribution will provide an adequate model for the hourly sales of newspapers throughout a day.

Sections 4.5 and 4.6

Test yourself   ANSWERS

1 (a) 0.966;   **(b)** 0.0867;   **(c)** 0.8795;   **(d)** 0.0338.

2 In theory there is no upper limit.

3 0.264.

4 Constant mean rate; cross independently.

5 (a) The average rate will not be constant over a 24-hour period;

(b) Cycles will obstruct each other so the crossings will not be independent.

6 (a) Mean 43.0, variance 1783.7;

(b) Large difference between mean and variance;

(c) Mean unlikely to be constant. More papers sold as people travel to and from work in morning and evening than in the middle of the day.

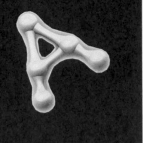

CHAPTER 5

Interpretation of data

Learning objectives

After studying this chapter, you should be able to:
- select a suitable diagram to illustrate qualitative, discrete quantitative and continuous quantitative data
- construct pie charts, bar charts, line diagrams, histograms, box and whisker plots, and cumulative frequency diagrams
- interpret statistical diagrams.

5.1 Introduction

You will be asked to interpret data in this section of the syllabus. The data may be presented in the form of a table of secondary data, diagrams and/or summary statistics. The specification states that you may be asked to construct pie charts, line diagrams, box and whisker plots, cumulative frequency diagrams and scatter diagrams. Scatter diagrams were dealt with in S1 and most of the other diagrams you will have met at GCSE. Despite, or perhaps because of this, questions on diagrams are often poorly answered in AS examinations.

Scatter diagrams are dealt with in S1.

The first step in analysing data is usually to draw a diagram. In some cases this will be sufficient and mathematical analysis will be unnecessary. In other cases it will reveal unusual features of the data which may make mathematical analysis inappropriate. It is always wise to look at the data before undertaking calculations. This chapter reviews the diagrams you may meet in the examination and concludes with some worked examples mostly based on tables of secondary data.

5.2 Qualitative data

> You should normally use a pie chart or a bar chart to illustrate qualitative data.

The pie chart opposite illustrates the proportion of votes cast for each party at the 1983 general election.

It is easy to see that the Conservatives received a little less than half the votes, Labour received more than a quarter and the Alliance just under a quarter. Other parties received only a small proportion of the votes.

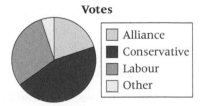

Votes
- Alliance
- Conservative
- Labour
- Other

Sometimes the number, or percentage, of votes for each party is written on the corresponding sector.

The diagram below places the seats won by the parties alongside the votes cast. It is easy to see that the Conservative won a bigger proportion of seats than they received votes, while the Alliance won a much smaller proportion of seats than votes.

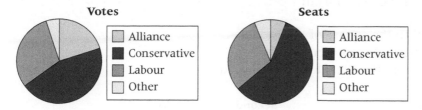

Drawing a pie chart

Roadside checks on the tyres on 300 cars in a suburban area produced the following data:

	Frequency	Proportion (freq/total)	Degrees (prop × 360)
Satisfactory	210	0.7	252
Slightly defective	55	0.1833	66
Seriously defective	35	0.1167	42
Total	**300**		

The angles subtended at the centre of the pie chart should be proportional to these frequencies. The proportion satisfactory is $\frac{210}{300} = 0.7$. Hence the satisfactory section should subtend an angle of $0.7 \times 360 = 252$ degrees at the centre.

A second set of roadside checks in a rural area produced the following results:

Satisfactory	110
Slightly defective	14
Seriously defective	23

You can compare the distributions by drawing two pie charts. The area of each chart should be made proportional to the total number of observations it represents. In this case there are 300 checks in the suburban area and 147 checks in the rural area.

Therefore the ratio of the areas should be $\frac{147}{300} = 0.49$.

If you choose to represent the suburban area by a circle of radius 2 cm, the area of the pie chart will be $\pi \times 2^2 = 4\pi$. If the circle representing the rural area has radius r its area will be πr^2.

$$\frac{\pi r^2}{4\pi} = 0.49, \quad r^2 = 4 \times 0.49 = 1.96, \quad r = 1.4 \text{ cm}$$

The area of a circle is proportional to the square of the radius. Whatever radius you choose for the circle representing the suburban area, the ratio of the radii of the two pie charts should be $\sqrt{0.49} = 0.7$.

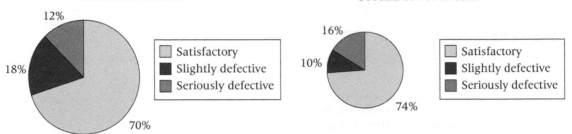

Bar charts

Four parties contested a parliamentary by-election.

The votes cast were as follows:

Labour	19 102
Conservative	18 329
Liberal Democrat	8196
Monster Raving Loony	672

If you drew a pie chart of this data it would not be easy to tell the difference between the Labour and Conservative section. As the most important thing about a by-election result is who got the most votes this information is best illustrated by a bar chart. The length of the bar is proportional to the number of votes received and it is clear to see that Labour got more votes than Conservative.

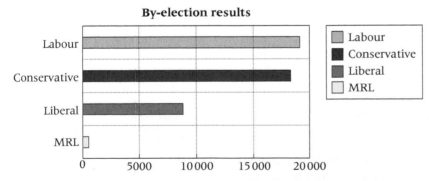

It is equally acceptable to draw the bars vertically instead of horizontally.

The party getting the most votes has been put at the top of the diagram followed by the other parties in order of votes obtained. This is not a compulsory feature of bar charts but is generally helpful in communicating the chart's message.

Worked example 5.1

An advertising campaign to promote electric showers consists of a mailshot which includes a pre-paid postcard requesting further details. Prospective customers who return the postcard are then contacted by one of five sales staff: Gideon, Magnus, Jemma, Pandora or Muruvet. The pie charts below represent the number of potential customers contacted and the number of sales completed during a 1-month period.

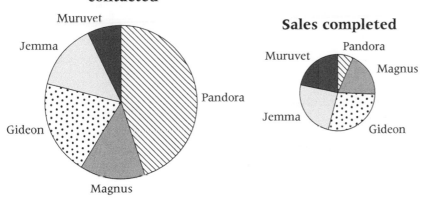

(a) The total number of potential customers contacted is 1100. Find, approximately, the total number of sales completed.

(b) Describe the main features of the data revealed by the pie charts.

(c) The manager wishes to compare the sales staff according to the number of sales completed. What type of diagram would you recommend, in place of a pie chart, so that this comparison could be made easily? [A]

Solution

(a) The radius of the **Potential customers** pie chart is approximately 2 cm while that of the **Sales completed** pie chart is approximately 1 cm.

$$\text{Total number of sales completed} = \frac{1100 \times 1^2}{2^2} = 275$$

(b) Pandora has contacted the most customers but completed the least sales. Muruvet has contacted the fewest customers but has made a similar number of sales to Jemma, Gideon and Magnus.

(c) Bar chart.

> You only need to comment on obvious differences. For example if you look carefully you can see that Gideon has made slightly more sales than Jemma. The difference is so small it does not require a comment.

5

EXERCISE 5A

1 The information below relates to people taking out mortgages. Draw an appropriate bar chart.

By type of dwelling (%)	
Type	All buyers
Bungalow	10
Detached house	19
Semi-detached house	31
Terraced house	31
Purpose built flat	7
Converted flat	3

2 The drinks purchased by customers at a cafe between 11.00 a.m. and midday were as follows:

coffee, coffee, tea, tea, tea, orange, tea, coffee, orange, orange, coffee, coffee, coffee, tea, tea, orange, coffee, coffee, orange, tea, tea, tea, tea.

Illustrate the data by means of a pie chart.

Between midday and 1.00 p.m. the cafe sold 24 coffees, 38 teas and nine oranges. Illustrate this data by means of a pie chart making the area of each pie chart proportional to the total frequency.

3 The following data from the 1991 census is the usual mode of travel to work for a 1% sample of people in employment:

None (home worker)	10 980
Train/tube/metro	13 456
Bus/coach	22 910
Car (driver)	124 293
Car (passenger)	18 106
Pedal cycle	6924
On foot	27 056
Other	9129

Illustrate the data:

(a) with a bar chart,

(b) with a pie chart.

Which features of the data are best illustrated by which diagram?

5.3 Discrete quantitative data

The number of times that a machine producing carpet tiles had to be adjusted on 11 successive night shifts is summarised below:

Number of adjustments	0	1	2	3
Frequency	3	2	5	1

Discrete quantitative data can be best illustrated by a line diagram. As with a bar chart the length of the line is proportional to the frequency.

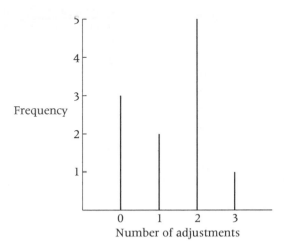

5.4 Continuous quantitative data

Continuous quantitative data can be best illustrated by histograms.

When yarn is delivered to a weaving mill its density is measured. The following data are the densities of 96 deliveries of yarn from the same supplier.

6300	6200	6150	6090	5720	5860	6220	6100
5910	5830	5950	6020	6120	6230	6090	6030
5910	5940	5930	5830	5890	5820	6560	5760
5900	5900	5820	6280	5740	6110	6200	6610
5960	6000	6070	5670	6400	6110	6290	6170
6440	6120	6120	6430	6220	6220	6370	6260
5560	6310	5950	5930	6780	6230	6340	6540
6320	5600	6170	6300	6260	5980	6740	5770
6310	6000	6430	6070	6150	6610	6310	6150
5630	6420	6020	6780	6200	6820	6470	6030
6110	6570	6150	6390	6650	6680	6620	6410
6370	6280	6480	5730	6280	5890	6230	6130

The units are denier which is the weight in grams of 9000 m of yarn. The reasons why this strange unit is used need not concern us.

Density is a continuous variable but the data recorded has been rounded to the nearest 10. Thus 6150 means that the actual density was between 6145 and 6155. The data may be formed into the frequency distribution below.

Class	Frequency	Interval width	Frequency density
5495–5595	1	100	0.01
5595–5695	3	100	0.03
5695–5795	5	100	0.05
5795–5895	7	100	0.07
5895–5995	11	100	0.11
5995–6095	10	100	0.10
6095–6195	14	100	0.14
6195–6295	15	100	0.15
6295–6395	10	100	0.10
6395–6495	8	100	0.08
6495–6595	3	100	0.03
6595–6695	5	100	0.05
6695–6795	3	100	0.03
6795–6895	1	100	0.01

These classes would not be satisfactory for some data. For example if the observation 5595 occurred you would not know whether to place it in the first or the second class. However, in this case the data was rounded to the nearest 10 and so an observation such as 5595 cannot occur. The classes are satisfactory for this data.

Histograms

This frequency distribution could be illustrated by a histogram. The area under the bar represents the frequency and the vertical axis represents frequency density. In this case the classes are all of equal width and so the height of the bars is proportional to the frequency.

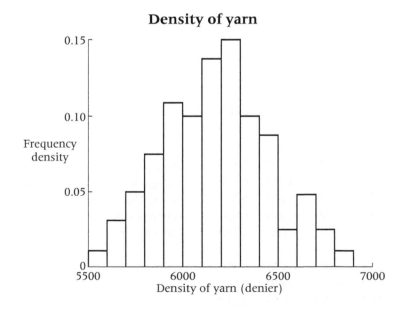

Histograms with classes of unequal width

If a frequency distribution has classes of unequal width the height of the bars in the histogram should **not** be proportional to the frequency. They should be proportional to the **frequency density**. The **frequency density** is the frequency divided by the class width.

Worked example 5.2

The table summarises the times between 90 consecutive admissions to an intensive care unit.

Time in hours	Frequency
0–	16
10–	22
30–	17
60–	15
100–200	19

Illustrate the data by means of a histogram. [A]

> This is an alternative way of showing classes. The first class contains all times between 0 and 10 hours.

Solution

Time in hours	Class width	Frequency	Frequency density
0–	10	16	1.6
10–	20	22	1.1
30–	30	17	0.567
60–	40	15	0.375
100–200	100	19	0.19

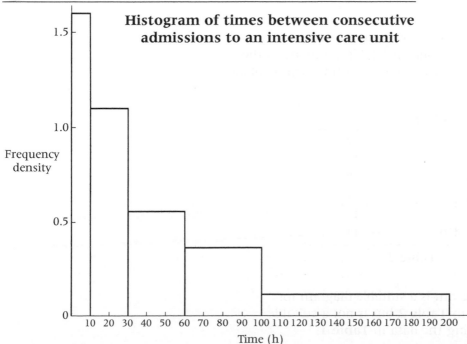

Histogram of times between consecutive admissions to an intensive care unit

Worked example 5.3

A travel agent carried out a survey among people who travel between Manchester and London at least 10 times per year.

Question 1 of the survey asked whether they made their last journey by car, coach, train or plane.

(a) State the type of diagram that would be most appropriate for illustrating the proportion of people using each method of transport.

Question 2 of the survey asked how many of the last four journeys were made by car. A summary of the results is shown below.

Number of journeys by car	Number of people
0	43
1	11
2	6
3	15
4	62

Table 1

(b) Illustrate the data by a line diagram. Comment briefly on the shape of the distribution.

Question 3 of the survey asked how long the last journey had taken from door to door. The following table summarises the journey times, in minutes, of those who had travelled by plane.

Time (minutes)	Frequency
180–	9
210–	11
240–	9
250–	11
260–	10
290–430	13

Table 2

(c) Explain why a histogram is a suitable diagram for illustrating the data in **Table 2** but is not a suitable diagram for illustrating the data in **Table 1**.

Solution

(a) Pie chart.

(b)

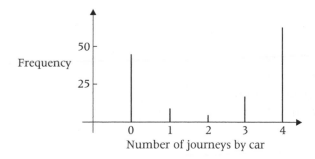

The distribution is U-shaped which is unusual.

(c) A histogram is suitable for the data in Table 2 as it is continuous. It is unsuitable for the data in Table 1 which is discrete quantitative.

EXERCISE 5B

1 The weight of a sample of expired blood donations were as follows:

Weight (g)	Frequency
70–79	8
80–89	24
90–99	26
100–109	12
110–119	5

Illustrate the data with a histogram.

2 The breaking strengths of 200 cables, manufactured by a company, are shown in the table below.

Breaking strength (in 100s of kg)	Frequency
0–	4
5–	48
10–	60
15–	48
20–	24
25–30	16

Draw a histogram of the data.

3

Time in minutes	Frequency
1–2	3
2–3	12
3–4	17
4–6	11
6–10	7

The distribution above refers to the time taken to complete a jigsaw puzzle by 50 people. Draw a histogram of the distribution.

4 The following table is extracted from a census report. It shows the age distribution of the population present on census night in Copeland, an area of Cumbria.

Population aged					
0–4	5–15	16–24	25–44	45–74	75 and over
4462	12 214	10 898	19 309	22 820	3364

Illustrate the data by means of a histogram. Make a suitable assumption about the upper bound of the class '75 and over'.

5 In an investigation children were asked a number of questions about their journeys to school. The data collected are to be illustrated using pie charts, line diagrams or histograms. (In answering this question you should only consider these three types of diagram.)

Question 1 asked the children whether they had travelled to school by bicycle, bus, train, car or on foot.

(a) Which type of diagram would best illustrate the proportion of children using each method of transport?

Question 2 asked the children on how many of the past 5 days they had travelled to school by car. A summary of the results is shown below.

Number of days by car	Number of people
0	112
1	32
2	18
3	7
4	29
5	64

(b) Illustrate these data using a suitable diagram. Comment briefly on the shape of the distribution.

Question 3 asked the children how long their journeys to school had taken on a particular day. A summary of the results is shown below.

Time (minutes)	Number of children
0.5–15.5	129
15.5–25.5	52
25.5–35.5	34
35.5–55.5	26
55.5–90.5	21

(c) Illustrate these data using an appropriate diagram.

Box and whisker plots

You will often wish to use a diagram to compare different sets of data. For example the weaving mill mentioned in section 5.4 received yarn from three suppliers. The densities illustrated there were measured on yarn from supplier A. Densities were also measured on samples of yarn from suppliers B and C. The results are shown below.

Densities of yarn from supplier B:

4240	4160	4120	4072	3776	3888	4176	4080
3928	3864	3960	4016	4096	4184	4072	4024
3928	3952	3944	3864	3912	3856	4448	3808
3920	3920	3856	4224	3792	4088	4160	4488
3968	4000	4056	3736	4320	4088	4232	5120
3100	3010	4096	4344	4176	4176	4296	4208
3648	4248	3960	3944	4624	4184	4272	4432
4256	3680	4136	4792	4208	3984	4592	3816
4248	4000	4344	4056	4120	4488	4248	4120
3704	4336	4016	4624	4160	4656	4376	4024
4088	4456	4120	4312	4520	4544	4496	4328
4296	4224	4384	3784	4224	3912	4184	4104

Densities of yarn from supplier C:

6166	5984	5893	5784	5110	5365	6020	5802
5456	5311	5529	5656	5838	6039	5784	5675
5456	5511	5493	5311	5420	5292	6639	5183
5438	5438	5292	6130	5147	5820	5984	6730
5547	5620	5747	5019	6348	5820	6148	5929
6421	5838	5838	6403	6020	6020	6293	6093
4819	6184	5529	5493	7040	6039	6239	6603
6202	4892	5929	7422	6093	5584	6967	5201
6184	5620	6403	5747	5893	6730	6184	5893
4947	6384	5656	7040	5984	7112	6475	7748
5820	6657	5893	6330	6803	6858	6748	6366
6239	6130	6494	5129	6130	5420	6039	5857

It is difficult to compare these distributions from the raw data. You could draw a histogram of each set of data. If the histograms were drawn on the same scale and placed one beneath the other a comparison could be made. An alternative method is to draw a box and whisker plot.

> This is particularly difficult to do if you are using a computer. Most common statistical packages are very poor at drawing histograms.

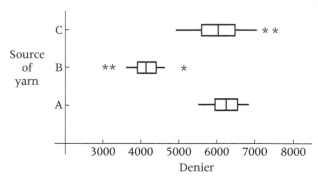

For each supplier the vertical line in the 'box' represents the median. The 'box' contains the middle 50% of the observations. The * represent **outliers** – that is observations which are untypical of the rest of the distribution. The 'whiskers' extending from the box join the box to the largest and smallest observations which are not outliers.

From the diagram it is easy to see that Supplier A had a slightly higher average density than Supplier C and that Supplier C was much more variable. Supplier B has a much lower average density than either of the others. The variability of Supplier B is similar to A apart from the three outliers.

> This is confirmed by numerical measures.
>
> | | Mean | Standard deviation |
> |---|---|---|
> | A | 6167 | 284 |
> | B | 4125 | 300 |
> | C | 5958 | 568 |

Constructing a box and whisker plot

The following data is the weight, in grams, of 16 pebbles sampled from a beach:

> 25.6 37.4 27.0 34.3 50.6 42.1 37.4 66.5
>
> 33.5 42.9 44.8 49.0 20.9 37.4 44.0 43.2

First you need to identify the median and the upper and lower quartiles. Arranging the data in order of magnitude:

> 20.9 25.6 27.0 33.5 34.3 37.4 37.4 37.4
>
> 42.1 42.9 43.2 44.0 44.8 49.0 50.6 66.5

There are 16 observations, so the median is halfway between the 8th and 9th, i.e. $\dfrac{(37.4 + 42.1)}{2} = 39.75$.

The lower quartile Q_1 is halfway between the 4th and 5th, i.e.

$$Q_1 = \frac{(33.5 + 34.3)}{2} = 33.9.$$

> See S1 chapter 2.

Similarly the upper quartile $Q_3 = \dfrac{(44.0 + 44.8)}{2} = 44.4$.

The generally accepted (but arbitrary) definition of an outlier is any observation:

> greater than $Q_3 + 1.5(Q_3 - Q_1)$
>
> or less than $Q_1 - 1.5(Q_3 - Q_1)$.

In this example an outlier is an observation:

> greater than $44.4 + 1.5(44.4 - 33.9) = 60.15$
>
> or less than $33.9 - 1.5(44.4 - 33.9) = 18.15$.

For this example the only outlier is 66.5.

Box and whisker plot of weights of pebbles

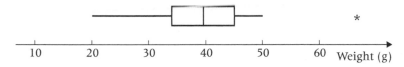

The lower whisker starts at 20.9, the lowest observation, but the upper whisker finishes at 50.6, since 66.5 is an outlier.

The left-hand whisker is longer than the right-hand whisker and the median is towards the upper end of the box. That is the smaller observations tend to be more spread out than the larger ones. If we ignore the outlier we would say that the distribution is **negatively skewed**.

Worked example 5.4

A factory manager obtains raw materials from three different suppliers. The time from placing the order to receiving the raw materials is called the **lead time**. For the factory to run smoothly, it is desirable that the distribution of this lead time should have a low average and low variability.

The diagram shows box and whisker plots of the lead times of recent deliveries of raw materials from suppliers A, B and C.

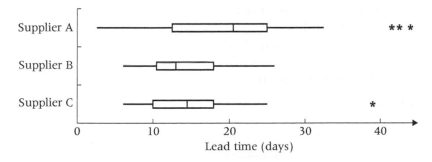

The lead times, in days, for the 12 most recent deliveries from Supplier C, as illustrated in the above diagram, were as follows:

> 7   9   6   12   11   15   17   14   15   25   19   39

(The lead times are in order, i.e. the most recent delivery had lead time 39 days, the second most recent 19 days, etc.)

(a) Explain, with the aid of suitable calculations, why, in the box and whisker plot for Supplier C, the lead time of 39 has been shown as an outlier, but the lead time of 25 has been included in the whisker.

(b) Compare, from the point of view of the factory manager, the lead times of the three suppliers as revealed by the box and whisker plots.

(c) What feature of the given data for Supplier C is not revealed by the box and whisker plot?

Solution

(a) For Supplier C, $Q_1 = 10$, $Q_3 = 18$ (from graph).
An observation $> 18 + 1.5(18 - 10) = 30$ is an outlier.

Hence 39 is an outlier.

A value of 25 is not greater than 30 but is greater than 18, the upper quartile. Hence it is included in the upper whisker.

(b) Supplier A's lead times are more variable and have a higher median than B or C. There are also three outliers representing very long lead times. Hence A is unsuitable. There is little to choose between B and C, but B has a slightly lower median and C has an undesirable outlier. Hence B would be preferred by the factory manager.

> Comment on the most obvious features. There is no need to go into great detail on, for example, the small differences in the length of the whiskers for Suppliers B and C.

(c) The box and whisker plot does not reveal that the lead times for C are tending to increase as time goes on.

In **(b)** you could also comment on the skewness. Ignoring the outliers, Supplier A is negatively skewed. This is a bad thing in this context as it means that there are more observations towards the upper end of the distribution (long lead times) than towards the lower end (short lead times). Supplier B is positively skewed which is a good thing in this context. Supplier C has a longer right-hand whisker but the median is towards the upper end of the box. For this supplier there is no useful comment to make on skewness.

EXERCISE 5C

1 Applicants for an assembly job are required to take a test of manual dexterity. The times, in seconds, taken to complete the task by 19 applicants were as follows:

63	229	165	77	49	74	67	59
66	102	81	72	59	74	61	82
48	70	86					

For these data find:

(a) the median,

(b) the upper and lower quartiles.

(c) Identify any outliers in the data.

(d) Illustrate the data by a box and whisker plot. Outliers, if any, should each be denoted by a * and should not be included in the whiskers.

2 The potencies of samples of two different brands of aspirin were as follows:

Brand A	58.7	58.4	59.3	60.4	59.8	59.4	57.7	60.3	61.0
	58.2	58.1	58.8	59.4	60.1	58.9			

Brand B	65.8	56.4	60.3	61.0	59.2	58.7	59.3	61.8	64.1
	61.9	62.8	60.7	55.2	59.9	63.6	60.4	59.9	61.5
	59.3	65.2	57.6	64.9	57.7	61.4	61.8	66.3	65.3

(a) Compare the potencies of the two brands by drawing box and whisker plots. You may use the following information about the sample of **Brand B**.

smallest value 55.2; lower quartile 59.3; median 61.0; upper quartile 63.6; largest value 66.3.

(Assume there are no outliers in either sample.)

(b) Comment, briefly, on the features of the data revealed by the diagrams.

3 A school cleaner is approaching pensionable age. She lives halfway between two post offices, A and B, and has to decide from which of the two she will arrange to collect her pension. For a few months she has deliberately used the two post offices alternately when she has required postal services. On each of these visits she has recorded the time taken between entering the post office and being served.

The box and whisker plots below show these waiting times for the two post offices. The symbol * represents an outlier.

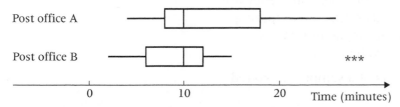

(a) Compare, in words, the distributions of the waiting times in the two post offices.

(b) Advise the cleaner which post office to use if the outliers were due to:

    (i) a cable-laying company having severed the electricity supply to the post office,

    (ii) the post office being short-staffed.

4 A group of athletes frequently run a cross-country course in training. The box and whisker plots below represent the times taken by athletes A, B, C and D to complete the course.

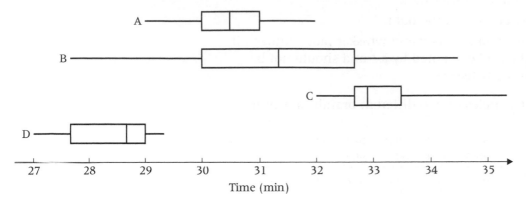

(a) Compare the times taken by athletes C and D.

Assume that the distributions shown above are representative of the times the athletes would take in a race over the same course.

(b) Which of the athletes A or B would you choose if you were asked to select one of them to win a race against:

(i) C,

(ii) D?

Give a reason for **each** answer.

(c) Which athlete would be most likely to win a race between A and B?

5.5 Cumulative frequency diagrams

Cumulative frequency diagrams provide a useful method of estimating the median and quartiles of grouped data. They are not a particularly good diagram for illustrating or comparing distributions. The cumulative frequency is the number of observations not greater than a given value.

The following table shows the weights of a sample of expired blood donations.

Weight (g)	Frequency	Cumulative frequency
70–79	8	8
80–89	24	32
90–99	26	58
100–109	12	70
110–119	6	76

These classes are suitable for data measured to the nearest gram. The class 70–79 will contain all blood donations weighing between 69.5 g and 79.5 g.

The cumulative frequency is plotted against the upper class bound. This is because we know there were eight samples weighing not more than 79.5 g, $8 + 24 = 32$ samples weighing not more than 89.5 g, $32 + 26 = 58$ samples weighing not more than 99.5 g, etc.

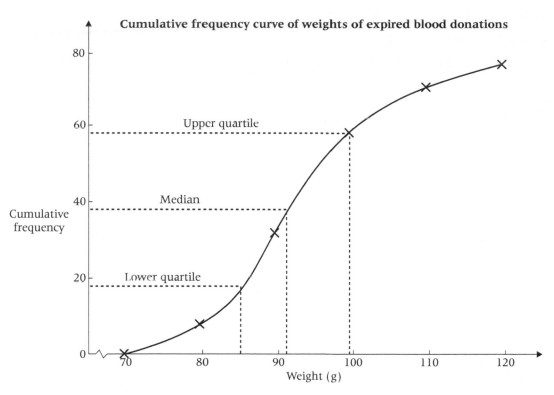

We may estimate the median of the sample by reading from the horizontal axis the value corresponding to a cumulative frequency of $\dfrac{(n + 1)}{2}$, where n is the sample size. The lower and upper quartiles are estimated by reading the value corresponding to $\dfrac{(n + 1)}{4}$ and $\dfrac{3(n + 1)}{4}$, respectively.

Unless n is small it makes little difference if you use $\dfrac{n}{4}, \dfrac{n}{2}$ and $\dfrac{3n}{4}$.

In this example $n = 76$ so we read off at $\dfrac{(76 + 1)}{2} = 38.5$ for the median and at $\dfrac{(76 + 1)}{4} = 19.25$ and $\dfrac{3(76 + 1)}{4} = 57.75$ for the quartiles. This gives an estimate of 91 for the median, 85 for the lower quartile and 99 for the upper quartile.

As we are estimating these values from grouped data, two significant figures is sufficient. To give more than two significant figures suggests greater accuracy than is possible in this case.

EXERCISE 5D

1 The lengths of 32 fish caught in a competition were measured correct to the nearest millimetre.

Length	20–22	23–25	26–28	29–31	32–34
Frequency	3	6	12	9	2

(a) Draw a cumulative frequency diagram.

(b) Estimate the median and quartiles of the lengths of the fish.

2 When laying pipes, engineers test the soil for 'resistivity'. If the reading is low then there is an increasing risk of pipes corroding. In a survey of 159 samples the following results were found:

Resistivity (ohms/cm)	Frequency
400–900	5
901–1500	9
1501–3500	40
3501–8000	45
8001–20 000	60

Draw a cumulative frequency diagram and estimate the median and quartiles.

3 The gross registered tonnage of 500 ships entering a small port are given in the following table.

Gross registered tonnage (tonnes)	No. of ships
0–	25
400–	31
800–	44
1200–	57
1600–	74
2000–	158
3000–	55
4000–	26
5000–	18
6000–8000	12

Plot the cumulative frequency diagram.

Hence estimate:

(a) the median tonnage,

(b) the interquartile range,

(c) the percentage of ships with a gross registered tonnage exceeding 2500 tonnes.

Worked example 5.5

The diagram compares the size of comprehensive schools with that of other public sector secondary schools in the United Kingdom.

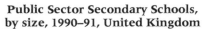

**Public Sector Secondary Schools,
by size, 1990–91, United Kingdom**

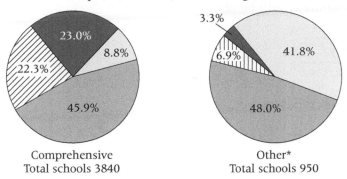

Comprehensive
Total schools 3840

Other*
Total schools 950

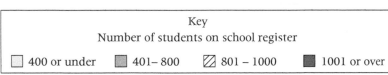

Key
Number of students on school register
☐ 400 or under   ▨ 401– 800   ▨ 801 – 1000   ■ 1001 or over

*Includes middle schools deemed secondary, secondary modern, grammar schools, technical and other schools.

Although this diagram is taken from official government statistics it is inappropriate to illustrate quantitative data with a pie chart.
In this question we extract the data from the pie chart and replace the pie chart with a more suitable diagram.

5

(a) How could the diagram be modified so that the number of schools in each category could also be compared?

You are not required to redraw the diagram.

(b) Using the percentages and total given in the diagram, or otherwise, construct a grouped frequency table for the size of **comprehensive schools**. (Size is measured by the number of students on the school register.)

For **comprehensive schools** the number of students on the school register is 210 in the smallest school and 1600 in the largest.

(c) (i) Draw a cumulative frequency diagram for the size of **comprehensive schools** and use it to estimate the upper and lower quartiles and the median.

(ii) Draw a box and whisker plot for the size of **comprehensive schools**.

(iii) Draw, on the same axes, a box and whisker plot for the size of the **other schools** given the following data.

Smallest	Lower quartile	Median	Upper quartile	Largest
88	290	468	694	1095

(d) Compare briefly, the distributions of size of comprehensive schools and other schools. Which type of diagram is more helpful in making this comparison?

Solution

(a) The area of the pie chart could be made proportional to the number of schools. The ratio of the radii would be

$$\sqrt{\frac{950}{3840}} = 0.497,$$

i.e. radius of the 'other schools' pie chart would be about half that of the 'comprehensive schools'.

(b)

Size of school	Frequency
400 or under	$0.088 \times 3840 = 338$
401–800	$0.459 \times 3840 = 1763$
801–1000	$0.223 \times 3840 = 856$
1001 and over	$0.230 \times 3840 = 883$

(c) (i)

Size of school	Cumulative frequency
400 or under	338
401–800	2101
801–1000	2957
1001 and over	3840

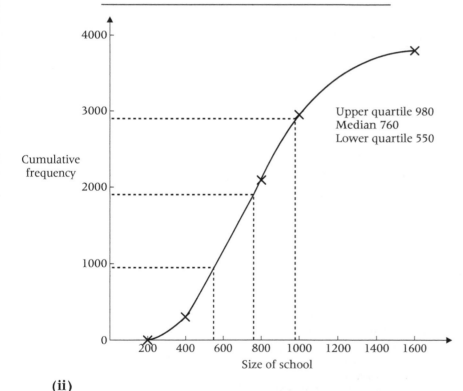

Upper quartile 980
Median 760
Lower quartile 550

You also need to use the additional information that the smallest school has 210 students and the largest 1600.

In this case $n = 3840$.

It makes no difference whether you use $\frac{n}{2}$ or $\frac{n+1}{2}$ for the median.

Read off at $\frac{3}{4} \times 3840 = 2880$ for the upper quartile.

$\frac{3840}{2} = 1920$ for the median.

$\frac{3840}{4} = 960$ for the lower quartile.

(ii)

Comprehensive

(iii) Other

There are no outliers.

(d) Comprehensive schools are larger, on average and more variable than other schools.

This is clear from the box and whisker plot, but much more difficult to see from the pie chart.

5.6 Interpretation of data

Questions on this section of the syllabus will be either based on a table of secondary data or on a real application of statistics. There are no specific techniques to learn although you may require knowledge of topics from S1 as well as SS02. This section consists of worked examples.

Worked example 5.6

The table below is copied from the *Annual Abstract of Statistics 2000*, Office for National Statistics.

London Underground: receipts, operations and assets

	Unit	1988 /89	1989 /90	1990 /91	1991 /92	1992 /93	1993 /94	1994 /95	1995 /96	1996 /97	1997 /98	1998 /99
Receipts												
Passenger: total	£ million	432	462	531	559	589	637	718	765	797	899	977
Ordinary*	,,	228	262	295	307	322	350	396	430	449	510	547
Season tickets	,,	204	200	236	252	267	287	322	335	348	389	430
Traffic												
Passenger journeys: total	Million	815	765	775	751	728	735	764	784	772	832	866
Ordinary	,,	363	380	399	368	365	376	398	416	418	448	463
Season tickets	,,	452	385	376	383	363	359	366	368	354	384	403
Passenger kilometres	,,	6 292	6 016	6 164	5 895	5 758	5 814	6 051	6 337	6 153	6 479	6 716
Operations												
Loaded train kilometres	Million	50.5	50.1	52.4	52.5	52.5	52.7	54.8	57.2	58.6	62.1	61.2
Place kilometres	,,	43.6	43.0	44.9	45.2	45.3	45.6	49.4	51.6	52.2	55.5	54.8
Rolling stock												
Railway cars	Number	3 950	3 908	3 880	3 880	3 895	3 955	3 923	3 923	3 867	3 886	3 923
Seating capacity	Thousand	169.8	171.6	171.6	166.9	166.9	168.5	167.7	165.4	164.5	164.0	165.0
Permanent way and stations												
Route kilometres open for traffic	Kilometres	394	394	394	394	394	394	392	392	392	392	392
Stations	Number	246	245	246	246	246	245	245	245	245	245	245

*Includes one day travelcards and concessionary fares.

Source: Department of the Environment, Transport and the Regions.

(a) What were the London Underground receipts from season tickets in 1997/98?

(b) What were the average receipts per passenger kilometre in 1988/89 and in 1998/99?

(c) What was the average cost of a journey for passengers using ordinary (i.e. not season) tickets in 1988/89 and in 1998/99.

(d) Comment on your results in **(c)** given that the United Kingdom Retail Price Index was 110.3 in December 1988 and 164.4 in December 1998.

Solution

(a) £389 million.

(b) Average receipts per passenger kilometre

1988/89 $\dfrac{432 \times 10^6}{6292 \times 10^6} = 0.0687$

   6.87p per passenger km

1998/99 $\dfrac{977 \times 10^6}{6716 \times 10^6} = 0.1455$

   14.55p per passenger km.

(c) Average cost of journey using ordinary ticket

1988/89 $\dfrac{228 \times 10^6}{363 \times 10^6} = 0.628$

   62.8p per journey

1998/99 $\dfrac{547 \times 10^6}{463 \times 10^6} = 1.18$

   £1.18 per journey

(d) Retail price index has risen from 110.3 to 164.4

   $\dfrac{164.4}{110.3} = 1.49.$ RPI has risen 49%

Average cost of journey has risen from 62.8p to £1.18,

   $\dfrac{118}{62.8} = 1.88.$ Cost risen by 88%

Rise in average cost of journey far exceeds rise in RPI.

> Be careful to include the units.
>
> A ruler will help you to identify the required figure accurately.

> Here 10^6 cancels but this will not always be the case. See Worked example 5.3**(d)**.

Worked example 5.7 ————————————

The following table is copied from the *Annual Abstract of Statistics*, 1999, ONS.

United Kingdom airlines*
Operations and traffic on scheduled services: revenue traffic

	Unit	1988	1989	1990	1991	1992	1993	1994	1995	1996	1997	1998
Domestic services												
Aircraft stage flights:												
Number	Number	277 682	303 147	300 683	285 346	299 893	300 416	301 652	318 884	331 109	336 218	352 936
Average length	Kilometres	279.0	281.0	288.0	301.0	305.7	311.3	314.8	317.0	320.0	330.0	333.1
Aircraft-kilometres flown	Millions	77.4	85.2	86.5	86.0	91.6	93.5	94.9	101.1	105.8	111.0	
Passengers uplifted	„	11.2	12.2	12.7	11.6	11.6	12.1	13.0	14.0	15.0	15.9	16.6
Seat-kilometres used	„	4 381.1	4 767.6	5 020.8	4 663.7	4 728.2	4 933.8	5 334.0	5 753.6	6 204.3	6 745.7	6 947.5
Tonne-kilometres used:	Millions											
Passenger	„	355.2	392.0	412.0	382.3	387.4	405.2	417.3	485.0	527.8	568.9	592.6
Freight	„	10.1	8.9	8.7	6.7	6.6	5.6	6.3	6.9	7.4	6.1	6.0
Mail	„	6.3	7.1	7.6	7.4	7.0	6.5	6.7	6.6	6.4	6.0	6.0
Total	„	371.5	408.0	428.3	396.5	401.1	417.3	430.3	498.5	541.6	581.0	604.7
International services												
Aircraft stage flights:												
Number	Number	268 704	303 358	316 794	282 776	301 607	301 204	319 620	339 714	371 400	413 588	444 746
Average length	Kilometres	1 361	1 320	1 381	1 456	1 523	1 629	1 693	1 703	1 695	1 688	1 729
Aircraft-kilometres flown	Millions	365.8	400.5	437.4	411.7	459.5	490.8	541.4	578.8	629.5	698.2	768.8
Passengers uplifted	„	20.2	22.9	25.7	22.9	26.5	28.0	30.9	33.5	36.1	40.4	45.1
Seat-kilometres used	„	59 487.1	65 428.3	74 558.8	69 951.7	82 003.1	89 736.3	99.0	109.6	118.6	129.7	145.0
Tonne-kilometres used:	Millions											
Passenger	„	5 514.6	6 212.9	7 053.6	6 625.6	7 747.8	8 500.1	9 352.3	10 636.4	11 661.9	12 718.2	14 162.3
Freight	„	2 047.2	2 197.4	2 380.1	2 373.2	2 637.4	2 914.0	3 371.8	3 560.4	3 824.5	4 448.0	4 657.2
Mail	„	172.9	155.2	161.1	175.2	154.0	135.0	140.5	144.4	169.5	166.3	171.7
Total	„	7 734.8	8 565.6	9 594.7	9 174.0	10 539.3	11 549.1	12 864.6	14 391.2	15 655.9	17 332.5	18 991.2

*Includes services of British Airways and other UK private companies.

Source: Civil Aviation Authority.

(a) How many passengers were uplifted by United Kingdom airlines domestic services in 1993?

(b) Describe the trend exhibited by freight tonne-kilometres for:

  (i) domestic services,

  (ii) international services.

Compare the two series.

(c) The aircraft-kilometres flown for domestic services in 1998 has been obliterated from the table. Use other data from the table to calculate this figure. Indicate how you have calculated your answer.

Solution

(a) 12.1 million.

(b) **(i)** There is a downward trend from 1988 to 1992. After that there is no clear trend, just apparently random variability.

(ii) There is a clear upward trend in this series. This is approximately linear although there is some indication that the rate of increase may be accelerating.

The freight tonne-kilometres for international services is much larger than that for domestic services. It starts about 200 times as big and the differing trends lead to this ratio increasing rapidly.

(c) Aircraft-kilometres flown =

(aircraft stage flights) $\times$ (average length)

For domestic services in 1998 = $352\,936 \times 333.1$

$= 117\,562\,982$

$= 117.6$ million

Worked example 5.8

The following table contains data on student loans in the UK in 1990/91 and in 1998/99.

	1990/91			1998/99		
	Number of loans (thousand)	Total sum borrowed (£million)	Average size of loan (£)	Number of loans (thousand)	Total sum borrowed (£million)	Average size of loan (£)
Male						
Living at home	6	2	300	48	77	1600
Away from home (London)	14	6	430	34	77	2230
Away from home (Other)	85	33	390	237	444	1870
All male	105	41	390	320	598	1870
Female						
Living at home	3	1	300	45	72	
Away from home (London)	10	4	430	39	88	2230
Away from home (Other)	57	22	390	255	476	1870
All female	71	27	390	340	635	1870

(a) What was the total sum borrowed in 1998/99 by females living at home?

(b) If a pie chart of the number of loans taken out by females in 1990/91 was drawn:

(i) what angle would be subtended at the centre by the sector representing females 'living at home',

(ii) what would be its radius, given that the radius of a similar pie chart for 1998/99 was 4 cm?

(c) The total number of loans taken out by all males in 1998/99 is shown as 320 000. The total is made up of three categories 'living at home', 'away from home (London)' and 'away from home (other)'. The sum of the three numbers shown for these categories is $48 + 34 + 237 = 319$ thousand. Does this indicate that there must be an error in the data? Explain your answer.

(d) The average size of loan (£) for females living at home in 1998/99 is not shown. Calculate this value.

(e) Describe, briefly, the main features revealed by the table. Try to make four different points.

Solution

(a) £72 million.

(b) (i) There were 71 000 loans to females in 1990/91 of which 3000 were to females living at home. Angle subtended at the centre of a pie chart by this sector is

$$\left(\frac{3}{71}\right) \times 360 = 15.2 \text{ degrees.}$$

(ii) The pie chart for 1998/99 represents 340 000 loans.

The ratio of the areas is therefore $\dfrac{71}{340} = 0.2088$.

The radius of the 1990/91 chart is therefore:

$$\sqrt{0.2088} \times 4 = 1.8 \text{ cm.}$$

(c) This does not imply an error. It is probably simply due to rounding each of the categories to the nearest thousand.

(d) $\dfrac{£72\,000\,000}{45\,000} = £1600.$

(e) Big increase in number of loans from 1990/91 to 1998/99.

Big increase in average size of loans.

In 1990/91 many more males than females took loans. By 1998/99 more females than males took loans.

That is, the average size of loans for males and females is nearly the same in all categories.

You may find other points to make.

Worked example 5.9

The table below is copied from the *Annual Abstract of Statistics*, 1999, ONS.

Motor vehicle production[†]
United Kingdom

Number

	1988	1989	1990	1991	1992	1993	1994	1995	1996	1997	1998
Motor vehicles											
SIC 1992, Class 34-10											
Passenger cars: total	1 226 835	1 299 082	1 295 610	1 236 900	1 291 880	1 375 524	1 466 823	1 532 084	1 686 134	1 698 001	1 748 258
1000 cc and under	129 446	133 135	93 039	15 918	22 037	98 034	98 178	95 198	108 645	119 894	112 044
Over 1000 cc but not over 1600 cc	764 289	716 784	809 219	496 822	793 307	709 615	729 397	814 873	845 084	829 086	814 595
Over 1600 cc but not over 2800 cc	260 231	375 309	325 116	193 972	437 951	515 487	573 357	528 444	635 861	653 154	720 556
Over 2800 cc	72 869	73 854	68 236	26 001	38 585	52 388	65 891	93 569	96 544	95 881	101 063
Commercial vehicles: total	317 343	326 590	270 346	217 141	248 453	193 467	227 815	233 001	238 314	237 703	227 379
Of which:											
Light commercial vehicles	250 053	267 135	230 510	105 633	216 477	171 141	197 285	199 346	205 372	210 942	203 629
Trucks:											
Under 7.5 tonnes	19 732	17 687	10 515	5 379	9 558	4 755	8 154	9 523	9 812	6 254	5 006
Over 7.5 tonnes	24 887	21 083	13 674	7 673	11 113	8 269	10 016	11 717	9 229	7 930	7 002
Motive units for articulated vehicles	6 171	5 827	3 327	1 444	2 788	2 283	2 794	3 476	208	2 573	2 492
Buses, coaches and mini buses	16 500	14 858	12 320	5 593	8 517	7 019	9 566	8 939	9 254	10 004	9 250

[†]Figures relate to periods of 52 weeks (53 weeks in 1988 and 1993).

Source: Office for National Statistics: 01633 812963.

(a) How many trucks over 7.5 tonnes were produced in the United Kingdom in 1995?

(b) How many motor vehicles in total were produced in the United Kingdom in 1996?

(c) The total number of passenger cars produced in 1998 is shown as 1 748 258. Comment on the number of significant figures given.

(d) Explain why the figures given for 1991 must contain some errors.

(e) The footnote suggests that the figures for 1988 and 1993 are compiled on a slightly different basis than the other years. Suggest a method of modifying the figures for 1988 and 1993 to give a fairer comparison.

(f) Describe, briefly, the overall trend shown by the total number of passenger cars produced in the United Kingdom and compare this with the trend shown by the total number of commercial vehicles produced.

Solution

(a) 11 717.

(b) 1 686 134 passenger cars + 238 314 commercial
vehicles = 1 924 448 motor vehicles.

(c) The seven significant figures given are excessive. It is highly
unlikely that this number could be correct to 7 significant
figures or that anyone would require this figure to such a
level of accuracy.

(d) In 1991 (unlike other years) the totals are not equal to the
sum of the constituent parts. For example for passenger
cars:

$$1\,236\,900 \neq 15\,918 + 496\,822 + 193\,972 + 26\,001.$$

(e) Figures for 1988 and 1993 could be multiplied by $\dfrac{52}{53}$.

(f) Total passenger car production shows a fairly steady upward
trend apart from a dip around 1990 to 1992.

Commercial vehicle production fell sharply but erratically
from 1989 to 1993 but was fairly constant at around
230 000 from 1994 to 1998.

Worked example 5.10

The following table is copied from the *Annual Abstract of Statistics
2000*, ONS.

Duration of unemployment in the United Kingdom

Thousands, Spring each year, not seasonally adjusted

Year	All unemployed	Less than 3 months	3 months & less than 6 months	6 months & less than 1 year	1 year & less than 2 years	2 years & less than 3 years	3 years & less than 4 years	4 years & less than 5 years	5 years or more	All 1 year or more Number	As % of total
1989	2075	647	306	333	252	133	103	71	229	788	38.0
1990	1974	686	324	310	211	107	73	51	210	653	33.1
1991	2414	834	466	434	276	113	67	41	179	676	28.0
1992	2769	668	500	607	529	174	75	37	179	993	35.9
1993	2936	600	474	599	612	287	109	57	196	1262	43.0
1994	2736	609	388	488	514	310	166	81	179	1249	45.7
1995	2454	568	386	422	404	243	143	102	182	1074	43.8
1996	2334	600	381	419	344	189	128	85	185	931	39.9
1997	2034	599	317	326	288	148	83	72	197	789	38.8
1998	1766	592	325	263	217	109	68	42	148	584	33.1
1999	1741	620	326	276	209	87	45	39	138	518	29.7

Source: Labour Force Survey, ONS

(a) How many people in the United Kingdom had been
unemployed for 5 years or more in 1990?

(b) How many people in the United Kingdom had been
unemployed for 3 years or more in 1998?

(c) Draw a histogram showing the duration of unemployment in the United Kingdom in 1999. Omit the category '5 years or more'.

(d) The bar chart shows for each of the years 1989 to 1999:
- the total unemployed men;
- men unemployed for 1 year or more;
- the total unemployed women;
- women unemployed for 1 year or more.

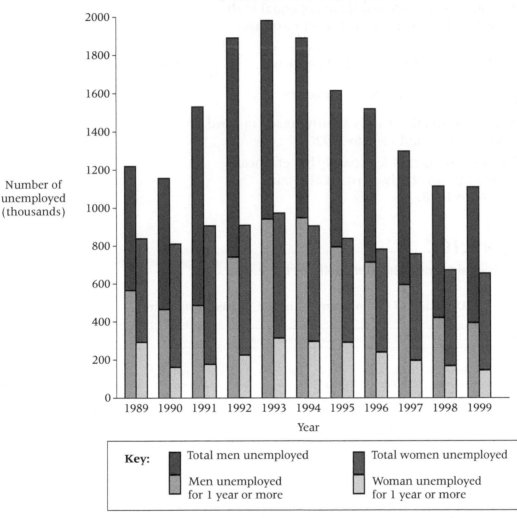

Number of unemployed men and women in the United Kingdom

Comment on the unemployment of men and women over the period 1989 to 1999, making **four** distinct points.

Solution

(a) 210 000

(b) 68 + 42 + 148 = 258 thousand
= 258 000

(c)

	frequency, thousands	frequency density
<3 months	620	2480
3–6 months	326	1304
6–12 months	276	552
1–2 years	209	209
2–3 years	87	87
3–4 years	45	45
4–5 years	39	39

Width of class '<3 months' is 3 months $\approx \frac{1}{4}$ year.
$620 \div \frac{1}{4} = 2480$.

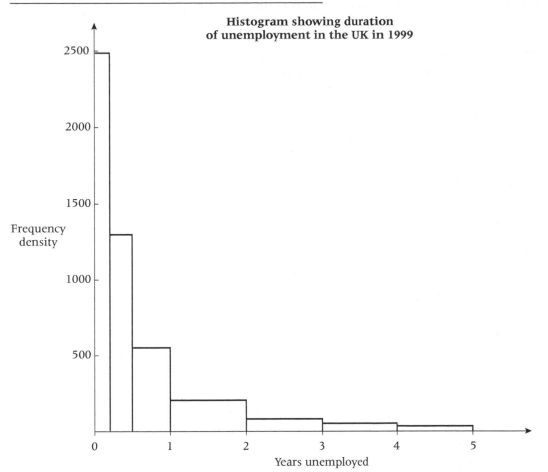

Histogram showing duration of unemployment in the UK in 1999

(d) Unemployment for both men and women rises to a peak in 1993 and then falls year by year to 1999.
More men than women unemployed in each year.
Ratio of men to women unemployed appears to increase as number unemployed increases.
Proportion of unemployed who have been unemployed for 1 year or more, fairly constant for both men and women.

There are many other possible answers.

Worked example 5.11

The following information refers to legislation for pre-packaged goods. It is extracted from the manual of 'Practical Guidance for Inspectors'.

Package – a container, together with the predetermined quantity of goods it contains, made up in the absence of the purchaser in such a way that none of the goods can be removed without opening the container.

Nominal quantity, Q_n – the quantity marked on the container.

Tolerable negative error, *TNE* – the negative error in relation to a particular nominal quantity as defined by the table below.

Tolerance limit, T_1 – the nominal quantity minus the tolerable negative error

$$T_1 = Q_n - TNE.$$

Absolute tolerance limit, T_2 – the nominal quantity minus twice the tolerable negative error

$$T_2 = Q_n - 2TNE.$$

Non-standard package – a package whose contents are less than the tolerance limit, T_1.

Inadequate package – a package whose contents are less than the absolute tolerance limit, T_2.

Tolerable negative errors (weight or volume)

Nominal quantity, Q_n (g or ml)	Tolerable negative error, *TNE*	
	% of Q_n	g or ml
5–50	9	–
50–100	–	4.5
100–200	4.5	–
200–300	–	9
300–500	3	–
500–1000	–	15
1000–10 000	1.5	–
10 000–15 000	–	150
Above 15 000	1	–

Thus, for tins of lentils with nominal quantity 400 g, the *TNE* is 12 g, and a non-standard tin is one containing less than 388 g.

Packets of peppermint tea have nominal quantity 40 g.

(a) Find:

    (i) the tolerable negative error,

    (ii) the tolerance limit,

    (iii) the absolute tolerance limit.

(b) If the contents of the packets are normally distributed with mean 40.5 g and standard deviation 2.5 g, find the proportion of packets which are:

(i) non-standard,

(ii) inadequate.

(c) Legislation requires that:
- condition I – the average contents of packages must be at least Q_n;
- condition II – not more than 2.5% of packages may be non-standard;
- condition III – not more than 0.1% of packages may be inadequate.

(i) Comment on the packets of peppermint tea in relation to these requirements.

(ii) Suggest a possible reason why it might be easier to meet these requirements for packets of tea with nominal quantity 50 g than for packets of tea with nominal quantity 90 g.

Solution

(a) (i) $TNE = \dfrac{9}{100} \times 40 = 3.6\,\text{g}$

(ii) $T_1 = 40 - 3.6 = 36.4\,\text{g}$

(iii) $T_2 = 40 - 2 \times 3.6 = 32.8\,\text{g}$

(b) (i) $z = \dfrac{36.4 - 40.5}{2.5} = -1.64$

Proportion non-standard
$= 1 - 0.949\,50 = 0.0505.$

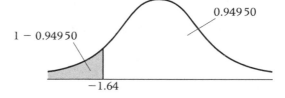

(ii) $z = \dfrac{32.8 - 40.5}{2.5} = -3.08$

Proportion inadequate
$= 1 - 0.998\,96 = 0.001\,04.$

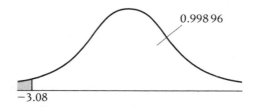

(c) (i) Condition I met.
Condition II not met.
Condition III not met but difference is negligible.
To meet condition II, either mean must be increased or standard deviation reduced.

(ii) *TNE* has same magnitude for $Q_n = 50$ and $Q_n = 90$.
However it is 9% for $Q_n = 50$ and only 5% for $Q_n = 90$.
Probably easier to ensure 97.5% contain at least 91% of Q_n than 97.5% contain at least 95% of Q_n.
Similarly for inadequate.

EXERCISE 5E

1 The weight of luggage belonging to each of the 135 passengers on an aeroplane is summarised in the frequency distribution shown:

Weight	0–	20–	30–	35–	40–	60–90
Number of passengers	23	28	31	4	33	16

(a) Illustrate these data using a histogram.

(b) Comment on the distribution of luggage weight given the fact that on this flight any passenger taking more than 35 kg of luggage had to pay an excess baggage charge.

(c) Draw a box-and-whisker plot of the data above using the following additional information.

Lightest luggage	Lower quartile	Median	Upper quartile	Heaviest luggage
4 kg	25 kg	32 kg	53 kg	87 kg

There are no outliers

(d) Discuss which of your two diagrams better illustrates the distribution of luggage weight.

2 Each member in a group of 100 children was asked to do a simple jigsaw puzzle. The times for the children to complete the jigsaw are summarised in the table below:

Time (seconds)	No. of children
60–85	7
90–105	13
110–125	25
130–145	28
150–165	20
170–185	5
190–215	2

(a) Illustrate the data with a cumulative frequency curve.

(b) Estimate the median and the inter-quartile range.

(c) Each member of a similar group of children completed a jigsaw in a median time of 158 seconds with an interquartile range of 204 seconds. Comment briefly on the relative difficulty of the two jigsaws.

In addition to the 100 children who completed the first jigsaw, a further 16 children attempted the jigsaw but gave up, having failed to complete it after 220 seconds.

(d) Estimate the median time taken by the whole group of 116 children.

Comment on the use of the median instead of the mean.

3 A company is considering three types of machine, *A*, *B*, and *C*, for producing tablets containing a certain drug. Samples of about 100 tablets produced by each type of machine were analysed. The box and whisker plots below illustrates the potency of the tablets (the symbol * is used to represent outliers).

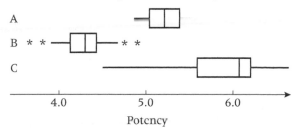

(a) Compare the distributions in words.

(b) Small variability in the potency of the tablets is desirable. The target value for each tablet is 6.0. It is easy to change the mean amount, but not possible to reduce the variability. Recommend, giving reasons, which type of machine should be used.

4 The pie charts below illustrate the sales of records, tapes, CDs and DVDs for two rival shops during May 2000. The area of each chart is proportional to the total number of sales.

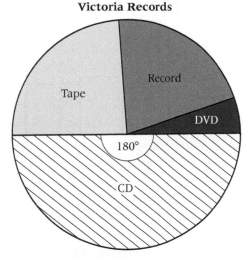

(a) During May 2000, Ron's Jazz Record Shop sold 760 CDs. Estimate the number of CDs sold by Victoria Records during the same period.

(b) Each shop would like to be able to have sold more DVDs than the other.

  (i) The pie charts above are unsuitable for showing which shop has sold more DVDs. Suggest a type of diagram which would enable this comparison to be made.

  (ii) Write down two facts about the sales of DVDs which can be seen easily from the pie charts.

5 Sally's Safaris is a holiday company which organises adventurous holidays. The ages, in years, of the customers who booked holidays in the year 2002 are summarised in the table below.

Age	Frequency
15–24	35
25–29	29
30–34	20
35–44	27
45–59	30
60–79	28

(a) Sally drew the histogram below to illustrate the data. Unfortunately, both coordinates of the point marked *R* have been plotted incorrectly.

State the correct coordinates of the point *R*.

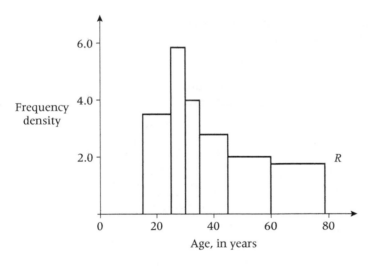

(b) Sally decides to draw pie charts for the years 2001 and 2002, illustrating the proportion of each year's customers in full-time employment.

   (i) Explain why histograms are unsuitable for this purpose.

   (ii) In 2002, 112 of the customers were in full-time employment. Calculate the angle at the centre of the pie chart for the sector representing these customers.

   (iii) The total number of customers in 2001 was 196. A pie chart of radius 1.4 cm was drawn. Calculate the appropriate radius for the 2002 pie chart, given that the areas of the pie charts are proportional to the total numbers of customers.

6 ## GB cinema exhibitor statistics

	Sites (number)	Screens (number)	Total no. of admissions (million)	Gross box office takings (£million)	Amount paid out for films (£million)	Revenue per admission (£)
1988	495	1 117	75.2	142.2	55.1	1.89
1989	481	1 177	82.9	169.5	65.4	2.04
1990	496	1 331	78.6	187.7	69.5	2.39
1991	537	1 544	88.9	229.7	78.5	2.58
1992	480	1 547	89.4	243.7	81.9	2.73
1993	495	1 591	99.3	271.3	95.9	2.73
1994	505	1 619	105.9	293.5	104.6	
1995	475	1 620	96.9	286.5	96.3	2.96
1996	483	1 738	118.7	373.5	144.9	3.15
1997	504	1 886	128.2	443.2	165.9	3.46
1998	481	1 975	123.4	449.5	162.6	3.64

Source: Office for National Statistics.

(a) How many admissions to GB cinemas were there in 1994?

(b) Describe the trend in the number of cinema:

    (i) sites,   **(ii)** screens.

Extend your answer to include comments on the number of screens per site.

(c) The figure for revenue per admission in 1994 has been obliterated. Use the data available to calculate this figure.

(d) Comment on the revenue per admission given that the Retail Price Index was 106.6 in June 1988 and 163.4 in June 1998.

7 ## Landings of fish by United Kingdom vessels: live weight and value into United Kingdom

	Quantity (thousand tonnes)							Value (£ thousand)					
	1992	1993	1994	1995	1996	1997		1992	1993	1994	1995	1996	1997
Brill	0.5	0.4	0.5	0.5	0.5	0.5		1 629	1 502	1 927	2 201	2 639	2 244
Catfish	1.9	1.7	1.4	1.0	1.1	1.0		1 878	1 867	1 669	1 474	1 449	1 152
Cod	62.2	64.8	65.7	74.4	75.7	71.0		71 570	64 915	65 090	65 772	69 752	67 641
Dogfish	14.1	10.4	8.2	10.9	9.7	8.6		10 032	7 327	6 291	7 684	7 002	5 586
Haddock	53.9	87.1	92.9	85.3	89.1	82.6		40 071	54 710	61 017	54 734	54 333	44 778
Hake	4.6	4.3	3.2	3.1	2.8	2.7		10 757	11 535	8 363	7 361	7 346	6 375
Lemon Soles	5.6	5.0	4.8	4.6	5.1	5.2		11 195	11 073	11 931	10 115	11 897	11 786
Ling	6.1	7.7	8.1	9.8	9.2	9.4		4 901	5 210	5 832	7 483	6 945	6 568
Megrims	4.1	4.3	4.7	5.1	6.0	6.0		8 309	8 543	9 210	9 227	10 430	9 467
Monks or Anglers	15.0	16.7	17.7	22.2	29.7	25.6		30 910	32 271	34 328	39 451	51 468	45 947
Plaice	24.1	19.4	17.5	15.5	12.5	12.9		24 281	21 499	19 937	17 573	16 013	15 484
Pollack (Lythe)	3.1	3.4	2.9	3.3	2.9	3.0		3 039	3 074	2 564	2 935	2 563	2 491
Saithe	13.0	12.2	12.5	12.8	13.3	12.3		5 707	4 671	4 875	5 625	5 674	4 894
Sand Eels	4.2	2.9	8.7	7.3	9.3	14.5		185	124	424	398	505	815
Skates and Rays	7.9	7.3	7.3	7.6	8.3	7.2		5 884	5 899	6 322	6 543	7 451	6 215
Soles	2.9	2.6	2.8	2.8	2.5	2.3		13 161	12 795	13 454	14 200	14 068	14 800
Turbot	0.8	0.9	1.0	0.8	0.8	0.7		3 874	4 521	5 238	4 667	5 006	4 263
Whiting	44.2	45.6	41.8	39.8	37.3	34.5		21 680	20 748	20 135	18 918	18 962	15 939
Whiting, Blue	6.5	2.3	1.9	3.4	3.5	12.4		333	137	90	180	190	693
Whitches	2.1	2.0	2.1	2.3	2.3	2.3		2 459	2 305	2 867	2 915	3 068	2 263
Other Demersal	8.6	8.9	8.9	12.9	13.3	13.2		7 034	8 730	10 709	17 028	16 614	16 000

Source: Office for National Statistics.

(a) What quantity of dogfish was landed in the United Kingdom in 1996?

(b) What was the value of the dogfish landed in the United Kingdom in 1996?

(c) Describe, briefly, the trend in the quantity of:

    (i) plaice landed,   **(ii)** sand eels landed.

(d) Calculate the average value of a tonne of haddock landed in:

    (i) 1992,   **(ii)** 1997.

(e) Calculate the average value of a tonne of brill landed in:

    (i) 1992,   **(ii)** 1997.

(f) Comment on the likely accuracy of your answers to **(e)**.

(g) Comment on your answers to **(d)** and **(e)** given that the Retail Price Index in June 1992 was 139.3 and in June 1997 was 157.5.

8 The following table is copied from the Digest of Environmental Statistics No 20, 1998.

Number of councils* and number of sites† participating in the bottle bank scheme: 1977–1996

Great Britain

	Number of councils	Number of sites
1977	6	17
1978	20	76
1979	33	143
1980	119	433
1981	158	614
1982	232	1 230
1983	286	1 758
1984	316	2 144
1985	341	2 470
1986	341	2 842
1987	389	3 138
1988	416	3 653
1989	431	4 330
1990	444	5 842
1991	448	7 155
1992	456	8 703
1993	457	10 965
1994	458	12 858
1995	459	14 300
1996	431‡	15 609

Source: British Glass Manufacturers Confederation.

* Includes Isle of Man, Guernsey and Isles of Scilly.

† Best estimate, as exact number of sites is not known.

‡ The lower figure is entirely the result of local government reorganisation, which has resulted in a smaller number of councils.

(a) Describe, briefly, the trend in the number of sites participating in bottle bank schemes.

(b) Describe, briefly, the trend in the number of councils participating in bottle bank schemes.

(c) Why would it be misleading to include the 1996 figure in a time series graph showing the number of councils participating in bottle bank schemes?

(d) From 1979 to 1985 there was a rapid increase in the number of councils participating. Why would it be impossible for this rate of increase to continue until 1995?

(e) Find the average number of sites per participating council in 1977, 1982, 1988 and 1995. Comment.

9 The following table is extracted from 'Health in England 1998: investigating the links between social inequalities and health'.

Cigarette smoking status by gross household income and sex
Adults aged 16 and over

Cigarette smoking status	Current cigarette smoker (Observed %*)	Ex-cigarette smoker (Observed %*)	Never smoked (Observed %*)	Base = 100%
Gross household income				
Men				
£20 000 or more	26	28	46	1093
£10 000–£19 999	31	30	39	683
£5 000–£9 999	31	40	29	440
Under £5 000	48	29	22	197
Total	29	30	40	2413
Women				
£20 000 or more	21	20	60	982
£10 000–£19 999	27	20	52	726
£5 000–£9 999	28	22	49	740
Under £5 000	32	22	46	545
Total	25	21	54	2993

* The percentages refer to the proportions of the subgroups reporting each smoking status, so do not add to 100.

(a) According to the table, what percentage of men from households with gross income between £10 000 and £19 999 are ex-cigarette smokers?

(b) What percentage of women from households with gross income between £5 000 and £9 999 are ex-cigarette smokers?

(c) Describe the main feature of the data for women who have never smoked.

(d) Describe the main feature of the data for men who have never smoked and compare this data with the data for women who have never smoked.

10 The following table is copied from the Annual Abstract of Statistics 1999.

Electricity: fuel used in generation – United Kingdom

Million tonnes of oil equivalent

	1988	1989	1990	1991	1992	1993	1994	1995	1996	1997	1998
All generating companies: total fuels	79.4	80.2	77.4	78.3	78.0	76.8	75.2	76.5	78.4	77.9	79.5
Coal	53.1	51.6	50.0	50.0	46.9	39.6	37.1	36.1	33.4	28.9	30.0
Oil	7.1	7.1	8.4	7.6	8.1	5.8	4.1	3.6	3.5		1.4
Gas	1.0	0.5	0.6	0.6	1.5	7.0	9.9	12.5	16.4	20.9	22.2
Nuclear	15.7	17.7	16.3	17.4	18.5	21.6	21.2	21.3	22.2	23.0	23.3
Hydro (natural flow)	0.4	0.4	0.4	0.4	0.5	0.4	0.4	0.5	0.3	0.4	0.4
Other fuels used by UK companies	1.0	1.7	1.8	0.9	1.1	1.0	1.1	1.2	1.2	1.4	1.2
Net imports	1.1	1.1	1.0	1.4	1.4	1.4	1.5	1.4	1.4	1.4	1.1

Source: Department of Trade and Industry.

(a) How much oil was used in electricity generation in the United Kingdom in 1994?

(b) Describe the trend in the use for electricity generation of:
   (i) coal, **(ii)** gas, **(iii)** nuclear.

(c) If a pie chart was drawn showing the fuels used in electricity generation in the United Kingdom in 1998, what would be the angle subtended at the centre by the sector representing oil?

   How would this sector compare with the sector representing oil in a similar pie chart drawn for 1988?

(d) The table shows that the use of hydro (natural flow) was the same in each year from 1988 to 1991 and then increased by 25% in 1992. Comment on this.

(e) How much oil was used in electricity generation in the United Kingdom in 1997?

11 This question relates to the legislation for prepackaged goods which is described in Worked example 5.11.
   Jars of coffee granules have a nominal quantity of 100 g.

(a) Find:
   (i) the tolerable negative error,
   (ii) the tolerance limit,
   (iii) the absolute tolerance limit.

(b) If the contents of the jars are normally distributed with mean 99 g and standard deviation 2 g, find the proportion of packets which are:

    (i) non-standard,   **(ii)** inadequate.

(c) Comment on the jars of coffee in relation to the legal requirements.

(d) Packets of tea have nominal contents 75 g. They have mean contents of 76 g with a standard deviation of 4 g. Comment on these packets in relation to the legal requirements.

12   **International tourism**

	Visits to the UK by overseas residents (Thousands)	Spending in the UK by overseas residents (£ million current prices)	Spending in the UK by overseas residents (£ million constant 1995) prices	Visits overseas by UK residents (Thousands)	Spending overseas by UK residents (£ million current prices)	Spending overseas by UK residents (£ million constant 1995) prices
1988	15 799	6184	9142	28 828	8216	12 515
1989	17 338	6945	9567	31 030	9357	12 861
1990	18 013	7748	9853	31 150	9886	12 021
1991	17 125	7386	8627	30 808	9951	11 775
1992	18 536	7891	8784	33 836	11 243	12 678
1993	19 863	9487	10 188	36 720	12 972	13 184
1994	20 794	9786	10 050	39 630	14 365	14 852
1995	23 537	11 763	11 763	41 345	15 386	15 386
1996	25 163	12 290	11 954	42 050	16 223	15 897
1997	25 515	12 244	11 542	45 957	16 931	18 652
1998	25 745	12 671*	11 573*	50 872	19 489*	21 847*

Source: adapted from *Annual Abstract of Statistics,* (National Statistics © Crown Copyright 2001).

(a) How many visits to the UK were made by overseas residents in 1992?

(b) What was the average amount spent in the UK by overseas residents on a visit to the UK in 1993?

(c) The table shows spending in the UK by overseas residents both at current prices and at constant 1995 prices. Give one advantage and one disadvantage of using constant 1995 prices compared to using current prices.

(d) By examining the ratio of current prices to constant 1995 prices for the spending figures highlighted *, or otherwise, explain why there must be some typing errors in the table.

(e) Describe, briefly, four main features of the data over the period 1988 to 1998.

13 The following table is adapted from statistics published by the Department of Health.

General Dental Service: Selected statistics England

	1993–94	1994–95	1995–96	1996–97	1997–98	1998–99
Number of dentists (1)						
Total	15 773	15 885	15 951	16 336	16 728	17 245
of which principals (2)	15 143	15 084	15 064	15 280	15 509	15 820
Number of patients registered (thousands)						
Adults	21 530	21 050	19 994	19 524	19 383	16 721
Children	7396	7367	7292	7270	7367	6775
Number of courses of treatment (thousands)						
Adults	24 848	24 913	24 752	24 580	25 268	26 171
Expenditure (£ millions)						
Gross expenditure	1 221.7	1 279.4	1 289.5	1 323.1	1 347.6	1 437.7
Paid by patients	367.0	383.3	381.2	383.0	388.4	419.6
Paid out of public funds	854.7	896.1	908.4	940.1	959.1	1 081.1

1 Principals, assistants and vocational dental practitioners at 30 September.
2 A principal is general dental practitioner on a health authority/family health services authority list.

Source: adapted from *Health and Personal Social Services Statistics* (Government Statistical Service)

(a) How many children were registered in 1997–98?

(b) There were 862 assistants in 1998–99. How many vocational dental practitioners were there?

(c) Comment on the Expenditure paid out of public funds given that the average Retail Price Index was 141.9 for 1993–94 and 164.4 for 1998–99.

(d) Describe briefly **four** main features of the data over the period 1993 to 1999.

Key point summary	
1 Qualitative data may be illustrated by pie charts or bar charts. Pie charts illustrate proportions. Bar charts compare the numbers in each class.	*p76*
2 Discrete quantitative data can be illustrated by line diagrams.	*p81*
3 Continuous quantitative data should be illustrated by histograms.	*p81*
4 Cumulative frequency diagrams are useful for estimating the median and quartiles of grouped data.	*p93*

Test yourself	**What to review**
1 What type of data may be illustrated by a pie chart?	*Section 5.2*
2 A secretary arranges appointments for a company executive. He records the number of appointments on each working day. Why would a histogram be an unsuitable diagram for illustrating this data?	*Section 5.4*
3 Name a suitable type of diagram to illustrate the data in question **2**.	*Section 5.3*
4 Which numerical measures may be estimated from a cumulative frequency diagram?	*Section 5.5*
5 What, in general terms, is an outlier? In the context of box and whisker plots, what is the usual definition of an outlier?	*Section 5.4*
6 A pie chart of radius 5 cm illustrates the votes cast for each party at a local by-election. The total number of votes cast was 4000. A second pie chart is to be drawn illustrating the votes cast for each party at a parliamentary by-election. The total number of votes cast at this election was 32 500. Find an appropriate radius for this second pie chart.	*Section 5.2*
7 What does the vertical axis of a histogram represent?	*Section 5.4*
8 A student attempts to draw a histogram using frequency as the vertical scale. Under what circumstances would the resulting diagram be seriously misleading?	*Section 5.4*

5

Test yourself ANSWERS

1 Qualitative.

2 This data is discrete – a histogram illustrates continuous data.

3 Line diagram.

4 Median, quartiles, interquartile range.

5 An observation untypical of the distribution as a whole. The usual convention is to define an outlier as any observation less than $Q_1 - 1.5(Q_3 - Q_1)$ or greater than $Q_3 + 1.5 (Q_3 - Q_1)$, where Q_3 is the upper quartile and Q_1 is the lower quartile.

6 14.25 cm.

7 Frequency density.

8 If the classes were of unequal width.

CHAPTER 6

Hypothesis testing

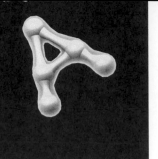

Learning objectives

After studying this chapter, you should be able to:

■ define a null and an alternative hypothesis

■ define the significance level of a hypothesis test

■ identify a critical region

■ understand whether to use a one- or a two-tailed test

■ understand what is meant by a **Type I** and a **Type II** error

■ test a hypothesis about a population mean based on a sample from a normal distribution with known standard deviation

> The population mean is denoted by μ.

■ test a hypothesis about a population mean based on a large sample from an unspecified population.

6.1 Forming a hypothesis

One of the most important applications of statistics is to use a *sample* to test an idea, or *hypothesis*, you have regarding a population.

Conclusions can never be absolutely certain but the risk of your conclusion being incorrect can be quantified (measured) and can enable you to identify *statistically significant* results.

> Statistically significant results require overwhelming evidence.

In any experiment, you will have your own idea or hypothesis as to how you expect the results to turn out.

A **Null Hypothesis,** written H_0, is set up at the start of any hypothesis test. This null hypothesis is a statement which defines the population and so always contains '=' signs, never '>', '<' or '≠'.

An example of a Null Hypothesis which you will meet in Worked example 6.1 of this chapter is:

H_0 Population mean lifetime of bulbs, $\mu = 500$ hours.

Usually, you are hoping to show that the Null Hypothesis is **not** true and so the **Alternative Hypothesis**, written H_1, is often the hypothesis you want to establish. Worked example 6.1 has:

H_1 Population mean lifetime of bulbs, $\mu > 500$ hours.

> The **null hypothesis** is only abandoned in the face of overwhelming evidence that it cannot explain the experimental results.
> Rather like in a court of law where the defendant is considered innocent until the evidence proves without doubt that he or she is guilty, the H_0 is accepted as true until test results show overwhelmingly that they cannot be explained if it was true.

> It often seems strange to students that they may want to show that H_0 is **not** true but, considering the examples of H_0 and H_1 given here, a manufacturer may well hope to show that bulbs have a **longer** than average lifetime.

A hypothesis test needs two hypotheses identified at the beginning: H_0 the **Null Hypothesis** and H_1 the **Alternative Hypothesis**.

H_0 states that a situation is unchanged, that a population parameter takes its usual value.
H_1 states that the parameter has increased, decreased or just changed.

6.2 One- and two-tailed tests

Tests which involve an H_1 with a $>$ or $<$ sign are called **one-tailed** tests because we are expecting to find just an increase or just a decrease.

Tests which involve an H_1 with a $\neq$ sign are called **two-tailed** tests as they consider any change (whether it be an increase or decrease).

For example, if data were collected on the amount of weekly pocket money given to a random selection of children aged between 12 and 14 in a rural area, and also in a city, it may be that you are interested in investigating whether children in the city are given **more** pocket money than children in rural areas. Therefore, you may set up your hypotheses as:

H_0 Population mean pocket money of children is the same in the rural area and in the city, or

$$\mu \, (\text{city}) = \mu \, (\text{rural})$$

H_1 Population mean pocket money **greater** in city, or

$$\mu \, (\text{city}) > \mu \, (\text{rural})$$

This is an example of a **one-tailed** test.

However, if you were monitoring the weight of items produced in a factory, it would be likely that **any** change, be it an increase or decrease, would be a problem and there would not necessarily be any reason to expect a change of a specific type.
In this case, typical hypotheses would be:

H_0 Population mean weight is 35 g

$$\mu = 35\text{g}$$

H_1 Population mean weight **is not** 35 g

$$\mu \neq 35\text{g}$$

This is an example of a **two-tailed** test.

One-tailed tests will generally involve words such as:
better or worse,
faster or slower,
more or less,
bigger or smaller,
increase or decrease.
In Worked example 6.1,
H_1 $\mu > 500$ hours indicates a **one-tailed** test.

Two-tailed tests will generally involve words such as:
different or difference,
change,
affected.

A **two-tailed** test is one where H_1 involves testing for any (non-directional) change in a parameter.
A **one-tailed** test is one where H_1 involves testing specifically for an increase or for a decrease (change in one direction only).

6.3 Testing a hypothesis about a population mean

Carrying out a hypothesis test to determine whether a population mean is significantly different from the suggested value stated in H_0 involves calculating a **test statistic** from a sample taken from the population.

As the test involves the population mean, it is the **sample mean**, $\bar{x}$, which must be evaluated. Since this test concerns a sample taken from a *normal* distribution, we also know that the sample means follow a normal distribution with mean equal to μ and with standard deviation equal to $\frac{\sigma}{\sqrt{n}}$.

The **test statistic** simply standardises the sample mean, $\bar{x}$, so that the result can be compared to critical z values.

It is very important that it can be assumed or known that the sample has been selected **randomly.** If the sample were not selected randomly, then valid conclusions regarding the whole population cannot be made since the sample may only represent one part of that population.

See S1, Chapter 5.

Test statistic $= \dfrac{\bar{x} - \mu}{\dfrac{\sigma}{\sqrt{n}}}$

6.4 The critical region and significance level of test

The **critical region** is the range of values of the test statistic which is so unlikely to occur when H_0 is true, that it will lead to the conclusion that H_0 is not true. The **significance level** of a test determines what is considered the level of overwhelming evidence necessary for the decision to conclude that H_0 is not true. It is the probability of wrongly rejecting a true H_0. The smaller the significance level, the more overwhelming the evidence required. Common values used for **significance levels** are 1%, 5% or 10%.

The test introduced in this chapter is based on a sample from a normal distribution. Therefore, the **critical region** is identified by finding critical z values from Table 4, 'Percentage

Table 4 is in the AQA Formulae book and in the Appendix.

points of the normal distribution', in exactly the same way as the z values were found in order for confidence intervals to be constructed.

See S1 chapter 6.

Some examples of critical regions are illustrated below.

One-tailed tests at 5% significance level

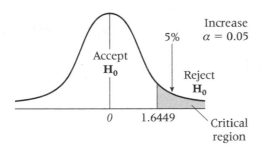

A 5% significance level is denoted $\alpha = 0.05$.

$z_\alpha = 1.6449$.

$z_\alpha = -1.6449$.

One-tailed tests at 1% significance level

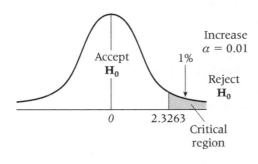

A 1% significance level is denoted $\alpha = 0.01$.

$z_\alpha = 2.3263$.

$z_\alpha = -2.3263$.

Two-tailed tests at 5% and 10% significance levels

Two critical values – change at both ends is considered.

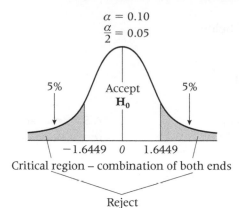

$\alpha = 0.10$
$\frac{\alpha}{2} = 0.05$

5% Accept 5%
$\mathbf{H_0}$

-1.6449 0 1.6449

Critical region – combination of both ends

Reject
$\mathbf{H_0}$

$z_{\frac{\alpha}{2}} = \pm 1.6449 \quad \alpha = 0.10.$

$\alpha = 0.05$
$\frac{\alpha}{2} = 0.025$

$2\frac{1}{2}\%$ Accept $2\frac{1}{2}\%$
$\mathbf{H_0}$

-1.96 0 1.96

Critical region – combination of both ends

Reject
$\mathbf{H_0}$

$z_{\frac{\alpha}{2}} = \pm 1.96 \quad \alpha = 0.05.$

The **critical region** or **critical value** identifies the range of extreme values which lead to the **rejection** of $\mathbf{H_0}$.

The **significance level**, α, of a test is the probability that a test statistic lies in the extreme critical region, if $\mathbf{H_0}$ is true. It determines the level of overwhelming evidence deemed necessary for the rejection of $\mathbf{H_0}$.

6.5 General procedure for carrying out a hypothesis test

The general procedure for hypothesis testing is:
1 Write down $\mathbf{H_0}$ and $\mathbf{H_1}$ 4 Identify the critical region
2 Decide which test to use   5 Calculate the test statistic
3 Decide on the significance level   6 Make your conclusion in context

Worked example 6.1

The lifetimes (hours) of Xtralong light bulbs are known to be normally distributed with a standard deviation of 90 hours.

A random sample of ten light bulbs is taken from a large batch produced in the Xtralong factory after an expensive machinery overhaul.

The lifetimes of these bulbs were measured as

523 556 678 429 558 498 399 515 555 699 hours.

Before the overhaul the mean life was 500 hours.

Investigate, at the 5% significance level, whether the mean life of Xtralong light bulbs has increased after the overhaul.

> The bulbs may appear to have a longer mean lifetime now but this may not be statistically significant.

Solution

The important facts to note are:

We are testing whether the mean is still 500 hours or whether an increase has occurred.

This means that $\mathbf{H_0}$ is $\mu = 500$ hours
   and $\mathbf{H_1}$ is $\mu > 500$ hours a **one-tailed** test.

The test is at the **5%** significance level.

The lifetimes are normally distributed with known standard deviation of 90 hours.

The hypothesis test is carried out as follows:

$\mathbf{H_0}$ $\mu = 500$ hours
$\mathbf{H_1}$ $\mu > 500$ hours $\alpha = 0.05$

From tables, the critical value is z = 1.6449 for this one-tailed test.

The sample mean, $\bar{x} = 541$ hours, from a sample with n = 10.

The population standard deviation is known, $\sigma = 90$ hours.

Therefore the test statistic

$$\frac{\bar{x} - \mu}{\frac{\sigma}{\sqrt{n}}} = \frac{541 - 500}{\frac{90}{\sqrt{10}}} = 1.44$$

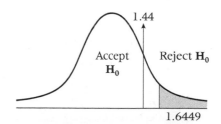

Conclusion

1.44 < 1.6449 for this one-tailed test, hence $\mathbf{H_0}$ is accepted.

There is no significant evidence to suggest that an increase in mean lifetime has occurred since the overhaul.

> Notice that we have not **proved** that $\mu = 500$ hours but we have shown that if $\mu = 500$ hours, a sample mean of 541 is not a particularly unlikely event. Conclusions should be interpreted in context.

Worked example 6.2

A forestry worker decided to keep records of the first year's growth of pine seedlings. Over several years, she found that the growth followed a normal distribution with a mean of 11.5 cm and a standard deviation of 2.5 cm.

Last year, she used an experimental soil preparation for the pine seedlings and the first year's growth of a sample of eight of the seedlings was

     7   22   19   15   11   18   17   15 cm.

Investigate, at the 1% significance level, whether there has been a change in the mean growth. Assume the standard deviation has not changed.

Solution

The important facts to note are:

We are testing whether the mean is 11.5 cm or not.

This means that H_0 is $\mu = 11.5$ cm
and H_1 is $\mu \neq 11.5$ cm a **two-tailed** test.

The test is at the **1%** significance level.

The growth is normally distributed with known standard deviation of 2.5 cm.

The hypothesis test is carried out as follows:

H_0 $\mu = 11.5$ cm
H_1 $\mu \neq 11.5$ cm $\alpha = 0.01$

From tables, the critical values are $z_{\frac{\alpha}{2}} = \pm 2.5758$ for this two-tailed test.

The sample mean, $\bar{x} = 15.5$, from a sample with $n = 8$.

The population standard deviation is known, $\sigma = 2.5$ cm.

Therefore the test statistic

$$\frac{\bar{x} - \mu}{\frac{\sigma}{\sqrt{n}}} = \frac{15.5 - 11.5}{\frac{2.5}{\sqrt{8}}} = 4.53$$

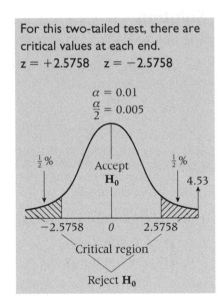

For this two-tailed test, there are critical values at each end.
$z = +2.5758$ $z = -2.5758$

Conclusion

$4.53 > 2.5758$ for this two-tailed test, hence H_0 is clearly rejected.

There is significant evidence to suggest that a change in mean growth has occurred. (It appears that the mean has increased.)

It is clear from the data that the mean has **increased**. You can conclude that the mean has **changed** or that it has **increased**. Either would be accepted in an examination. Conclusions should be interpreted in context

Worked example 6.3

The owner of a small vineyard has an old bottling machine which is used for filling bottles with his wine. The bottles contain a nominal 75 cl of wine.

The old machine is known to dispense volumes of wine which are normally distributed with mean 76.4 cl and a standard deviation of 0.9 cl.

The owner is concerned that his old machine is becoming unreliable and he decides to purchase a new bottling machine. The manufacturer assures the owner that the new machine will dispense volumes which are normally distributed with a standard deviation of 0.9 cl.

The owner wishes to reduce the mean volume dispensed.

A random sample of twelve 75 cl bottles are taken from a batch filled by the new machine and the volume of wine in each bottle is measured. The volumes were

    75.7   76.2   75.4   75.8   75.4   76.9
    76.4   75.5   76.1   76.8   76.7   76.5 cl.

Investigate, at the 5% significance level, whether the volume of wine dispensed has been reduced.

Solution

The important facts to note are:

We are testing whether the mean is still 76.4 cl or whether a decrease has occurred.

This means that H_0 is $\mu = 76.4$ cl
            and H_1 is $\mu < 76.4$ cl a **one-tailed** test.

The test is at the **5%** significance level.

The volumes are normally distributed with known standard deviation of 0.9 cl.

The hypothesis test is carried out as follows:

    $H_0 \; \mu = 76.4$ cl
    $H_1 \; \mu < 76.4$ cl $\alpha = 0.05$

From tables, the critical value is z $= -1.6449$ for this one-tailed test.

The sample mean, $\bar{x} = 76.11667$ cl, from a sample with n $= 12$.

The population standard deviation is known, $\sigma = 0.9$ cl.

Therefore the test statistic

$$\frac{\bar{x} - \mu}{\frac{\sigma}{\sqrt{n}}} = \frac{76.11667 - 76.4}{\frac{0.9}{\sqrt{12}}} = -1.09$$

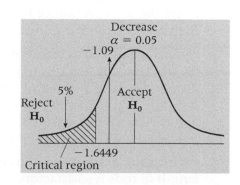

Conclusion

$-1.09 > -1.6449$ for this one-tailed test, hence H_0 is accepted.

There is no significant evidence to suggest that the mean volume has decreased.

EXERCISE 6A

1 A factory produces lengths of rope for use in boatyards. The breaking strengths of these lengths of rope follow a normal distribution with a standard deviation of 4 kg.

The breaking strengths, in kilograms, of a random sample of 14 lengths of rope were as follows:

    134   136   139   143   136   129   137
    130   138   134   145   141   136   139.

The lengths of rope are intended to have a breaking strength of 135 kg but the manufacturer claims that the mean breaking strength is in fact greater than 135 kg.

Investigate the manufacturer's claim using a 5% level of significance.

2 Reaction times of adults in a controlled laboratory experiment are normally distributed with a standard deviation of 5 s. Twenty-five adults were selected at random to take part in such an experiment and the following reaction times, in seconds, were recorded:

    6.5   3.4   5.6   6.9   7.1   4.9   10.9   7.8
    2.4   2.8   11.3   3.7   7.8   2.4   2.8   3.7
    4.9   12.0   6.5   12.8   6.9   7.4   3.1   1.9   11.5

Investigate, using a 5% significance level, the hypothesis that the mean reaction time of adults is 7.5 s.

3 A maze is devised and after many trials, it is found that the length of time taken by adults to solve the maze is normally distributed with mean 7.4 s and standard deviation 2.2 s. A group of nine children was randomly selected and asked to attempt the maze. Their times, in seconds, to completion were:

    6.1   9.0   8.3   9.4   5.8   8.1   7.6   9.2   10.0

Assuming that the times for children are also normally distributed with a standard deviation of 2.2 s, investigate, using the 5% significance level, whether children take longer than adults to do the maze.

4 A machine produces steel rods which are supposed to be of length 2 cm. The lengths of these rods are normally distributed with a standard deviation of 0.02 cm. A random sample of ten rods is taken from the production line and their lengths measured. The lengths are:

    1.99 1.98 1.96 1.97 1.99 1.96 2.0 1.97 1.95 2.01 cm.

Investigate, at the 1% significance level, whether the mean length of rods is satisfactory.

5 The resistances, in ohms, of pieces of silver wire follow a normal distribution with a standard deviation of 0.02 ohms. A random sample of nine pieces of wire are taken from a batch and their resistances were measured with results:

    1.53 1.48 1.51 1.48 1.54 1.52 1.54 1.49 1.51 ohms.

It is known that, if the wire is pure silver, the resistance should be 1.50 ohms but, if the wire is impure, the resistance will be increased.
Investigate, at the 5% significance level, whether the batch contains pure silver wire.

6 The weight of Venus chocolate bars is normally distributed with a mean of 30 g and a standard deviation of 3.5 g. A random sample of 20 Venus bars was taken from a new production line and was found to have a mean of 32.5 g. Is there evidence, at the 1% significance level, that the mean weight of bars from the new production line exceeds 30 g?

7 The weights of components produced by a certain machine are normally distributed with mean 15.4 g and standard deviation 2.3 g.
The setting on the machine is altered and, following this, a random sample of 81 components is found to have a mean weight of 15.0 g.
Does this provide evidence, at the 5% level, of a reduction in the mean weight of components produced by the machine? Assume that the standard deviation remains unaltered.

8 The ability to withstand pain is known to vary from individual to individual.
In a standard test, a tiny electric shock is applied to the finger until a tingling sensation is felt. When this test was applied to a random sample of ten adults, the times recorded, in seconds, before they experienced a tingling sensation were:

    4.2 4.5 3.9 4.4 4.1 4.5 3.7 4.8 4.2 4.2

Test, at the 5% level, the hypothesis that the mean time before an adult would experience a tingling sensation is 4.0 s. The times are known to be normally distributed with a standard deviation of 0.2 s.

6.6 Hypothesis test for means based on a large sample from an unspecified distribution

There are occasionally real life situations where we may wish to carry out a hypothesis test for a mean by examining a sample taken from a population where the standard deviation is known. However, it is much more likely that, if the mean of the population is unknown, the standard deviation will also be unknown. Provided that a large sample is available, then a sufficiently good estimate of the population standard deviation can be found from the sample. The use of a large sample also has the advantage that the sample mean is approximately normally distributed regardless of the distribution of the population.

As in section 6.5, the **sample mean**, $\bar{x}$, is evaluated and, since this test involves a **large** sample, we know that the sample mean is approximately normally distributed. The standard deviation is also evaluated and used as an estimate of σ for this test.

The **test statistic** is $\dfrac{\bar{x} - \mu}{\frac{\sigma}{\sqrt{n}}}$ as before, and is compared to critical z values.

> The definition of large is arbitrary but a sample size of $n \geqslant 30$ is usually considered 'large'.

> It is still important to ensure that the sample is randomly selected.

> As the sample is large it makes very little difference whether the divisor n or $n - 1$ is used when estimating σ. However as σ is being estimated from a sample it is correct to use the divisor $n - 1$.

To carry out a hypothesis test for a mean based on a **large** sample from an **unspecified** distribution:

the **test statistic** is $\dfrac{\bar{x} - \mu}{\frac{\sigma}{\sqrt{n}}}$.

An estimate of the standard deviation, σ, can be made from the sample, the **test statistic** is compared to **critical z values**. These are found in AQA Table 4.

Worked example 6.4

A manufacturer claims that the mean lifetime of her batteries is 425 hours. A competitor tests a random sample of 250 of these batteries and the mean lifetime is found to be 408 hours, with a standard deviation of 68 hours.

Investigate the claim of the competitor that the batteries have a mean lifetime less than 425 hours. Use a 1% significance level.

Solution

The important facts to note are:

We are testing whether the population mean lifetime is 425 hours or whether it is less than 425 hours.

This means that $\mathbf{H_0}$ is $\mu = 425$ hours
and $\mathbf{H_1}$ is $\mu < 425$ hours a **one-tailed** test.

The test is at the **1%** significance level.

The lifetimes are from an unspecified distribution with an unknown standard deviation. However, the sample size is **large** so $\bar{x}$ is approximately normally distributed and an estimate of the unknown population standard deviation can be calculated from the sample.

> $n = 250$.

The hypothesis test is carried out as follows:

H₀ $\mu = 425$ hours
H₁ $\mu < 425$ hours $\alpha = 0.01$

From tables, the critical value is z = −2.3263 for this one-tailed test.

The sample mean, $\bar{x} = 408$ hours, from a sample with n = 250.

The population standard deviation is estimated as $\sigma = 68$ hours.

The **test statistic** is $\dfrac{\bar{x} - \mu}{\dfrac{\sigma}{\sqrt{n}}} = \dfrac{408 - 425}{\dfrac{68}{\sqrt{250}}} = -3.95.$

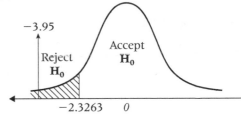

Conclusion

−3.95 < −2.3263 for this one-tailed test. Hence **H₀** is rejected.

There is significant evidence to suggest that the mean lifetime is less than the 425 hours claimed by the manufacturer.

Worked example 6.5

A precision machine is set to produce metal rods which have a mean length of 2 mm. A sample of 150 of these rods is randomly selected from the production of this machine. The sample mean is 1.97 mm and the standard deviation is 0.28 mm.

(a) Investigate, using the 5% significance level, the claim that the mean length of rods produced by the machine is 2 mm.

(b) How would your conclusions be affected if you later discovered that:

(i) the sample was not random,

(ii) the distribution was not normal?   [A]

Solution

(a) The important facts to note are:

We are testing whether or not the population mean length is 2 mm.

This is a **two-tailed** test.
The test is at the **5%** significance level.
The lengths are from an unspecified distribution with an unknown standard deviation. However, the sample size is **large**.
The hypothesis test is carried out as follows:

H₀ $\mu = 2$ mm
H₁ $\mu \neq 2$ mm $\alpha = 0.05$

From tables, the critical value is z = ±1.96 for this two-tailed test.

The sample mean, $\bar{x} = 1.97$ mm, from a sample with n = 150. The population standard deviation is estimated as $\sigma = 0.28$ mm.

The **test statistic** is $\dfrac{\bar{x} - \mu}{\frac{\sigma}{\sqrt{n}}} = \dfrac{1.97 - 2}{\frac{0.28}{\sqrt{150}}} = -1.31.$

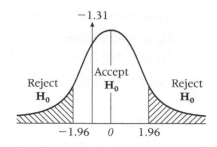

Conclusion

$-1.31 > -1.96$ for this two-tailed test. Hence **H_0** is accepted.

There is no significant evidence to suggest that a change in mean length has occurred.

> We have shown that if $\mu = 2$ mm, a sample mean of 1.97 is not a particularly unlikely event.

(b) **(i)** If the sample was not random there can be no confidence in the conclusion. For example, these may have been the first 150 rods produced and the mean length could have changed as production continued.

   (ii) It makes no difference whether the distribution is normal or not as this is a large sample.

EXERCISE 6B

1 A factory produces lengths of cable for use in winches. These lengths are intended to have a mean breaking strength of 195 kg but the factory claims that the mean breaking strength is in fact greater than 195 kg.
In order to investigate this claim, a random sample of 80 lengths of cable is checked. The breaking strengths of these sample lengths are found to have a mean of 199.7 kg and a standard deviation of 17.4 kg. Carry out a suitable hypothesis test, using a 1% level of significance, to test the factory's claim.

2 The resistances, in ohms, of 5 mm pieces of wire for use in an electronics factory are supposed to have a mean of 1.5. A random sample of 85 pieces of wire are taken from a large delivery and their resistances are carefully measured. Their mean resistance is 1.6 ohms and their standard deviation is 0.9 ohms. Investigate, at the 5% significance level, whether the mean resistance of all wires from the delivery is 1.5 ohms.

3 A Government department states that the mean score of Year 2 children in a new national assessment test is 78%.
A large education authority selected 90 Year 2 children at random from all those who took this test in their area and found that the mean score for these children was 72.8%, with a standard deviation of 18.5%.
Investigate, using a 5% significance level, the hypothesis that the mean score of children in this area is lower than 78%.

4 A large dairy company produces whipped spread which is packaged into 500 g plastic tubs. Sixty tubs were randomly selected from the production line and weighed. The mean weight of these tubs was found to be 504 g with a standard deviation of 17.3 g.

   Investigate, using the 5% significance level, whether the mean weight of the tubs is greater than 500 g.

5 A machine produces steel rods which are supposed to be of length 30 mm. A customer suspects that the mean length of the rods is less than 30 mm. The customer selects 50 rods from a large batch and measures their lengths.

   The mean length of the rods in this sample was found to be 29.1 mm with a standard deviation of 2.9 mm.

   Investigate, at the 1% significance level, whether the customer's suspicions are correct.

   What assumption was it necessary to make in order to carry out this test?

6 The weight of Bubbles biscuit bars is meant to have a mean of 35 g.

   A random sample of 120 Bubbles bars is taken from the production line and it is found that their mean weight is 36.5 g. The standard deviation of the sample weights is 8.8 g. Is there evidence, at the 1% significance level, that the mean weight is greater than 35 g?

7 Varicoceles is a medical condition in adolescent boys which may lead to infertility. A recent edition of *The Lancet* reported a study from Italy which suggested that the presence of this condition may be detected using a surgical instrument. The instrument gives a mean reading of 7.4 for adolescent boys who do not suffer from varicoceles. A sample of 73 adolescent boys who suffered from varicoceles gave a mean reading of 6.7 and a standard deviation of 1.2.

   (a) Stating your null and alternative hypotheses and using a 5% significance level, investigate whether the mean reading for all adolescent boys who suffer from varicoceles is less than 7.4.

   (b) Making reference to the value of your test statistic, comment briefly on the strength of the conclusion you have drawn.

   (c) State, and discuss the validity of, any assumptions you have made in (a) about the method of sampling and about the distribution from which the data were drawn. [A]

6.7 Significance levels and problems to consider

You may have wondered how the **significance level** used in hypothesis testing is chosen. You have read that significance levels commonly used are 1%, 5% or 10% but no explanation has been offered about why this is so.

A common question asked by students is:

> *Why is the level of overwhelming evidence necessary to lead to rejection of H_0 commonly set at 5% ?*

The **significance level** of a hypothesis test gives the
P(test statistic lies inside critical region | H_0 true).
In other words, *if* H_0 is true, then, with a 5% significance level, you would expect a result as extreme as this only once in every 20 times. If the test statistic does lie in the critical region the result is statistically significant at the 5% level and we conclude that H_0 is **untrue**.

Sometimes it may be necessary to be 'more certain' of a conclusion. If a traditional trusted piece of research is to be challenged, then a 1% level of significance may be used to ensure greater confidence in rejecting H_0.
If a new drug is to be used in preference to a well-known one then a 0.1% level may be necessary to ensure that no chance or fluke effects occur in research which leads to conclusions which may affect human health.

6.8 Errors

It is often quite a surprising concept for students to realise that, having correctly carried out a hypothesis test on carefully collected data and having made the relevant conclusion to accept the H_0 as true or to reject it as false, this conclusion might be right or it might be wrong.
However, you can never be absolutely certain that your conclusion is correct and has not occurred because of a *freak* result.
The significance level identifies for you the risk of a freak result leading to a wrong decision to reject H_0.
This leads many students to ask why tests so often use a **5% significance level** which has a probability of 0.05 of incorrectly rejecting H_0 when it actually is true. Why not reduce the significance level to 0.1% and then there would be a negligible risk of 0.001 of this error occurring?

The answer to this question comes from considering the **two** errors which may occur when conducting a hypothesis test.

This table illustrates the problems:

		Conclusion	
		H₀ true	**H₀** not true
Reality	**H₀** correct	Conclusion correct	Error made **Type I**
	H₀ incorrect	Error made **Type II**	Conclusion correct

The table shows that the **significance level** of a test is
P(conclusion **H₀** not true | **H₀** really is correct) =
P(**Type I** error made).

The other error to consider is when a test does not show a
significant result even though the **H₀** actually is **not** true.
P(conclusion **H₀** true | **H₀** really is incorrect)
= P(**Type II** error made).

The probability of making a **Type II** error is difficult or
impossible to evaluate unless precise further information is
available about values of the population parameters. If a value
suggested in **H₀** is only slightly incorrect then there may be a
very high probability of making a **Type II** error. If the value is
completely incorrect then the probability of a **Type II** error will
be very small.

Obviously, if you set a very low **significance level** for a test,
then the probability of making a **Type I** error will be low but
you may well have quite a high probability of making a **Type II**
error.

There is no logical reason why 5% is used, rather than 4% or 6%.
However, practical experience over a long period of time has
shown that, in most circumstances, a significance level of 5%
gives a good balance between the risks of making **Type I** and
Type II errors.

This is why 5% is chosen as the 'standard' significance level for
hypothesis testing and careful consideration must be given
before changing this value.

> Errors which can occur are:
> A **Type I** error which is to reject **H₀** when it is true.
> A **Type II** error which is to accept **H₀** when it is not true.

Not only can you conclude **H₀** is
true when really it is false but
also you could conclude it is false
when actually it is true.

You will not be expected to
evaluate the probability of
making a **Type II** error in SS02.

If a low risk of wrongly rejecting
H₀ is set, then it is unlikely that
the test statistic will lie in the
critical region. **H₀** is unlikely to
be rejected unless the null
hypothesis is 'miles away' from
reality.

Worked example 6.6

A set of times, measured to one-hundredth of a second, were
obtained from nine randomly selected subjects taking part in a
psychology experiment.

The mean of the nine sample times was found to be 9.17 s.

It is known that these times are normally distributed with a standard deviation of 4.25 s.

The hypothesis that the mean of such times is equal to 7.50 s is to be tested, with an alternative hypothesis that the mean is greater than 7.50 s, using a 5% level of significance.

Explain, in the context of this situation, the meaning of:

(a) a **Type I** error,
(b) a **Type II** error.

Solution

For this example:

H_0 $\mu = 7.50$ s
H_1 $\mu > 7.50$ s $\quad \alpha = 0.05$ one-tailed test.

(a) A **Type I** error is to reject H_0 and conclude that the population mean time is **greater than** 7.50 s when, in reality, the mean time for such an experiment is equal to 7.50 s.

> If H_0 is untrue it is impossible to make a **Type 1** error.

The probability of this happening, if H_0 is true is $\alpha = 0.05$.

> If H_0 is true it is impossible to make a **Type 2** error.

(b) A **Type II** error is to accept H_0 and conclude that the population mean time is **equal to** 7.50 s when, in reality, the mean time for such an experiment is greater than 7.50 s.

The probability of this happening will vary and can only be determined if more information is given regarding the exact alternative value that μ may take, not simply that $\mu > 7.50$ s.

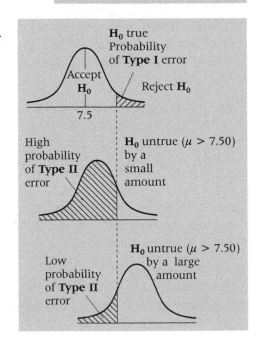

Worked example 6.7

Explain, in the context of Worked example 6.2, the meaning of:

(a) a **Type I** error,
(b) a **Type II** error.

Solution

In Worked example 6.2, we have:

H_0 $\mu = 11.5$ cm
H_1 $\mu \neq 11.5$ cm $\quad \alpha = 0.01$

Hence:

(a) A **Type I** error is to reject H_0 and conclude that the mean growth of seedlings is **not equal** to 11.5 cm when, in reality, the mean growth for seedlings grown in the experimental soil preparation is equal to 11.5 cm.

The probability of this happening if H_0 is true is $\alpha = 0.01$.

(b) A **Type II** error is to accept H_0 and conclude that the mean growth of seedlings is **equal** to 11.5 cm when, in reality, the mean growth for the seedlings grown in the experimental soil preparation is not equal to 11.5 cm.

As seen in the previous example, the probability of this cannot be determined unless precise information is given regarding the alternative value taken by μ.

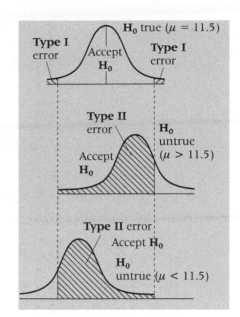

EXERCISE 6C

1 Refer to question **1** in Exercise 6A.

(a) Explain, in the context of this question, the meaning of:
  (i) a **Type I** error,
  (ii) a **Type II** error.

(b) What is the probability of making a **Type I** error in this question if:
  (i) H_0 is true,
  (ii) H_0 is untrue?

2 Refer to question **4** in Exercise 6A.

(a) Explain, in the context of this question, the meaning of:
  (i) a **Type I** error,
  (ii) a **Type II** error.

(b) What is the probability of making a **Type I** error in this question if:
  (i) H_0 is true,
  (ii) H_0 is untrue?

3 Times for glaze to set on pottery bowls follow a normal distribution with a standard deviation of three minutes. The mean time is believed to be 20 minutes.
A random sample of nine bowls is glazed with times which gave a sample mean of 18.2 minutes.

(a) Investigate, at the 5% significance level, whether the results of this sample support the belief that the mean glaze time is 20 minutes.

6

(b) Explain, in the context of this question, the meaning of a **Type I** error.

(c) Why is it not possible to find the probability that a **Type II** error is made?

4 The lengths of car components are normally distributed with a standard deviation of 0.45 mm. Ten components are selected at random from a large batch and their lengths were found to be,

19.3 20.5 18.1 18.3 17.6 19.0 20.1 19.2 18.6 19.4 mm.

(a) Investigate, at the 10% significance level, the claim that the mean length of such components is 19.25 mm.

(b) Explain, in the context of this question, the meaning of:

(i) a **Type I** error,

(ii) a **Type II** error.

(c) Write down the probability of making a **Type I** error if the mean length is 19.25 mm.

5 A random sample of ten assembly workers in a large factory are trained and then asked to assemble a new design of electrical appliance. The times to assemble this appliance are known to follow a normal distribution with a standard deviation of 12 minutes. It is claimed that the new design is easier to assemble and the mean time should be less than the mean of 47 minutes currently taken to assemble the old design of appliance. The mean time for the sample of ten workers was 39.8 minutes.

(a) Investigate, using a 5% significance level, the claim that the new design is easier to assemble.

(b) Explain, in the context of this question, the meaning of a **Type II** error.

Key point summary

1 A hypothesis test needs two hypotheses identified at the beginning: H_0 the **Null Hypothesis** and H_1 the **Alternative Hypothesis**. *p119*

H_0 and H_1 both refer to the population from which the sample is randomly taken.

2 H_0 states that a situation is unchanged, that a population parameter takes its usual value. *p119*

H_1 states that the parameter has increased, decreased or just changed.

H_0 states what is to be assumed true unless overwhelming evidence proves otherwise. In the case of testing a mean, H_0 is $\mu = k$ for some suggested value of k.

3 A **two-tailed** test is one where H_1 involves testing for any (non-directional) change in a parameter. *p120*

A **one-tailed** test is one where H_1 involves testing specifically for an increase or for a decrease (change in one direction only).

A **two-tailed test** results in a critical region with two areas.

A **one-tailed test** results in a critical region with one area.

4 The **critical region** or **critical value** identifies the range of extreme values which lead to the **rejection** of H_0. *p122*

The **critical values** for the tests introduced in this chapter can be found in Table 4 in the AQA Formulae Book.

5 The **significance level**, α, of a test is the probability that a test statistic lies in the extreme critical region, if H_0 is true. It determines the level of overwhelming evidence deemed necessary for the rejection of H_0. *p122*

The **significance level**, α, is commonly, but not exclusively, set at 1%, 5% or 10%.

6 The general procedure for hypothesis testing is: *p122*
1 Write down H_0 and H_1
2 Decide which test to use
3 Decide on the significance level
4 Identify the critical region
5 Calculate the test statistic
6 Make your conclusion in context.

7 The **test statistic** used for investigating a hypothesis regarding the mean of a normally distributed population with **known standard deviation**, σ, is *p120*

$$\frac{\bar{x} - \mu}{\frac{\sigma}{\sqrt{n}}},$$

where $\bar{x}$ is the mean of the randomly selected sample of size n.

8 The **test statistic** used for investigating a hypothesis regarding the population mean based on a **large** sample from an **unspecified distribution** is *p128*

$$\frac{\bar{x} - \mu}{\frac{\sigma}{\sqrt{n}}},$$

where σ can be estimated from the large sample.

If the test statistic lies **in** the **critical region**, or **beyond** the **critical value**, H_0 is rejected.

6

8 Errors which can occur are: *p133*
  A **Type I** error which is to reject H_0 when it is true.
  A **Type II** error which is to accept H_0 when it is
  not true.
  The probability of making a **Type I** error is usually
  denoted by α.

Test yourself	What to review
1 Which of the following hypotheses would require a one-tailed test and which a two-tailed test? **(a)** Amphetamines stimulate motor performance. The mean reaction time for those subjects who have taken amphetamine tablets will be faster than that for those who have not. **(b)** The mean score on a new aptitude test for a precision job is claimed to be lower than the mean of 43 found on the existing test. **(c)** Patients suffering from asthma have a higher mean health conscious index than people who do not suffer from asthma. **(d)** The mean length of rods has changed since the overhaul of a machine.	*Section 6.2*
2 What is the name given to the value with which a test statistic is compared in order to decide whether a null hypothesis should be rejected?	*Section 6.4*
3 (a) What is the name given to the agreed probability of wrongly rejecting a null hypothesis? **(b)** Give three commonly used levels for this probability.	*Section 6.4*
4 What is the rule which is commonly used to determine whether a sample is **large**?	*Section 6.6*
5 A manufacturer collects data on the annual maintenance costs for a random selection of eight new welding machines. The mean cost of these eight machines is found to be £54.36. The standard deviation of such costs for welding machines is £8.74. Stating the null and alternative hypotheses, and using a 1% significance level, test whether there is any evidence that the mean cost for maintenance of the new machines is less than the mean value of £71.90 found for the old welding machines.	*Section 6.5*

Test yourself (continued)	**What to review**

6 Carry out a hypothesis test at the 5% significance level and state your conclusion.

Section 6.6

Sample mean	$\bar{x} = 32.6$
Sample size	$n = 250$
Null hypothesis	$H_0: \mu = 34$
Alternative hypothesis	$H_1: \mu \neq 34$
Population standard deviation	unknown
Estimate of pop standard deviation from sample	17.6
Population distribution	unspecified

7 In a survey of workers who travel to work at a large factory by car, the distances, in kilometres, travelled by a random sample of ten workers were:

Sections 6.3, 6.5 and 6.8

   14   43   17   52   22   25   68   32   26   44

In previous surveys, the mean distance was found to be 35.6 km with a standard deviation of 14.5 km.

(a) Investigate, using a 5% significance level, whether the mean distance travelled to work has changed. Assume the standard deviation remains 14.5 km.

(b) What is the meaning of:
   (i) a **Type I** error,
   (ii) a **Type II** error,
   in the context of this question?

(c) Why is it important that the sample of workers is selected at random from all those factory workers who travel to work by car?

6

Test yourself **ANSWERS**

1 (a) one-tail; **(b)** one-tail; **(c)** one-tail; **(d)** two-tail.

2 Critical value.

3 (a) Significance level; **(b)** 1%, 5%, 10%.

4 n is at least 30.

5 $H_0: \mu = £71.90$; $H_1: \mu < £71.90$; ts -5.68 cv -2.3263 mean cost is less.

6 ts -1.26 cv ± 1.96 Accept H_0. No significant evidence to doubt $\mu = 34$.

7 (a) ts -0.284 cv ± 1.96 no change;
(b) (i) conclude there has been a change when in fact there has not;
(ii) conclude no change when in fact there has been a change;
(c) Conclusion unreliable if sample not random. For example, sample may have been taken only from white collar workers who may have a different mean travelling distance from manual workers.

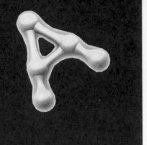

CHAPTER 7

Contingency tables

Learning objectives

After studying this chapter, you should be able to:

- analyse contingency tables using the χ^2 distribution
- recognise the conditions under which the analysis is valid
- combine classes in a contingency table to ensure the expected values are sufficiently large
- apply Yates' correction when analysing 2×2 contingency tables.

7.1 Contingency tables

A biology student observed snails on a bare limestone pavement and in a nearby limestone woodland. The colour of each snail was classified as light, medium or dark.

The following table shows the number of snails observed in each category.

	Light	Medium	Dark
Pavement	22	10	3
Woodland	8	10	12

A table, such as the one above, which shows the frequencies of two variables (colour and habitat) simultaneously is called a contingency table.

> It is possible for contingency tables to show the frequencies of more than two variables. In this module you will only meet tables showing two variables.

> A contingency table shows the frequencies of two (or more) variables simultaneously.

Contingency tables are analysed to test the null hypothesis that the two variables are independent. That is, in this case, that the proportion of light snails is the same in the woodland as on the limestone pavement as are the proportions of medium snails and of dark snails. Clearly the observed proportions are not the same – for example, a much larger proportion of light snails were observed on the pavement than in the woodland. However, (as in all hypothesis testing) the hypothesis refers to the population and the test is carried out to examine whether the observed sample could reasonably have occurred by chance if the null hypothesis was true.

To carry out the test, first calculate the expected number of snails you would observe in each category if the null hypothesis were true. To do this it is helpful to extend the table to include totals and sub-totals.

	Light	Medium	Dark	Total
Pavement	22	10	3	35
Woodland	8	10	12	30
Total	30	20	15	65

row totals

grand total

column totals

The table now shows the totals for each row and for each column (the sub-totals). It also shows the total number of snails observed.

If there are the same proportion of light snails on the pavement as in the woodland then the best estimate that can be made of this proportion is the total number of light snails observed divided by the total number of snails observed. That is $\frac{30}{65}$.

Since a total of 35 snails was observed on the pavement you would expect to observe $\left(\frac{30}{65}\right) \times 35 = 16.15$ light snails on the pavement.

The expected number refers to a long run average and so will usually not be a natural number.

Similarly you would expect $\left(\frac{30}{65}\right) \times 30$ light snails to have been observed in the woodland, $\left(\frac{20}{65}\right) \times 35$ medium snails to have been observed on the pavement, etc.

Notice that in each case the expected number in a particular cell is $\frac{(\text{row total}) \times (\text{column total})}{(\text{grand total})}$.

The expected number in any cell of a contingency table is $\frac{(\text{row total}) \times (\text{column total})}{(\text{grand total})}$.

This formula works for all contingency tables providing you are investigating the independence of two variables.

The following table shows the observed number, O, on the left of each cell and the expected number, E, on the right of each cell.

	Light	Medium	Dark	Total
Pavement	22, 16.15	10, 10.77	3, 8.08	35
Woodland	8, 13.85	10, 9.23	12, 6.92	30
Total	30	20	15	65

It is usually sufficient to calculate the *E*s to two decimal places. However the more significant figures used at this stage of the calculation the better.

Note that the total of the *E*s is the same as the total of the *O*s in each row and in each column. In this case it was only necessary to derive two Es – say for the expected number of light and medium snails observed on the pavement – and the rest could have been deduced from the totals.

The test statistic is $X^2 = \Sigma \dfrac{(O - E)^2}{E}$. This will have a small value if the observed frequencies in each cell are close to the frequencies expected. It will have a large value if there are big differences between the frequencies observed and those expected. Hence the null hypothesis will be accepted if $X^2 = \Sigma \dfrac{(O - E)^2}{E}$ is small and rejected if it is large.

> X^2 is used as it is similar but not identical to χ^2. There is no universally recognised symbol for this statistic.

The test statistic X^2 is approximately distributed as a χ^2 distribution provided the Os are frequencies (i.e. not lengths, weights, percentages, etc.) and the Es are reasonably large (say >5). Note the contingency table must also be complete. For example, it would not be permissible to leave the dark snails out of the analysis of the contingency table above.

> For the analysis to be valid it must be possible to allocate each snail examined to one and only one cell.

> $X^2 = \Sigma \dfrac{(O - E)^2}{E}$ may be approximated by the χ^2 distribution provided:
> **(i)** the Os are frequencies,
> **(ii)** the Es are reasonably large, say >5.

To obtain a critical value from the χ^2 distribution it is necessary to know the degrees of freedom. General rules exist for deriving degrees of freedom but in the case of an $m \times n$ contingency table all you need to know is that the number of degrees of freedom is $(m - 1)(n - 1)$.

> An $m \times n$ contingency table has $(m - 1)(n - 1)$ degrees of freedom.

The contingency table above is 2×3 and so there are

$$(2 - 1)(3 - 1) = 2 \text{ degrees of freedom.}$$

Note that 2 was also the number of Es which had to be derived from the null hypothesis before the rest could be calculated from the totals. This is not a coincidence and is one way of interpreting degrees of freedom.

> Alternatively, note that once the sub-totals are known there are only two independent frequencies. Once these are known the rest are fixed.

The analysis of the contingency table can now be completed.

	Light	Medium	Dark	Total
Pavement	22, 16.15	10, 10.77	3, 8.08	35
Woodland	8, 13.85	10, 9.23	12, 6.92	30
Total	30	20	15	65

H$_0$ Colour is independent of whether a snail is found on limestone pavement or in woodland.
H$_1$ Colour is not independent of whether a snail is found on limestone pavement or in woodland.

$$X^2 = \Sigma \frac{(O - E)^2}{E} = \frac{(22 - 16.15)^2}{16.15} + \frac{(10 - 10.77)^2}{10.77}$$

$$+ \frac{(3 - 8.08)^2}{8.08} + \frac{(8 - 13.85)^2}{13.85}$$

$$+ \frac{(10 - 9.23)^2}{9.23} + \frac{(12 - 6.92)^2}{6.92}$$

$$= 11.6$$

You may prefer to set the calculation out in a table

O	E	$\dfrac{(O - E)^2}{E}$
22	16.15	2.1190
10	10.77	0.0551
3	8.08	3.1939
8	13.85	2.4709
10	9.23	0.0642
12	6.92	3.7292

$$X^2 = \Sigma \frac{(O - E)^2}{E} = 11.6$$

The Os are frequencies and the Es are all greater than five and so we can compare the calculated value of X^2 with a critical value from the χ^2 distribution. The degrees of freedom as calculated above are $(2 - 1)(3 - 1) = 2$.

> The test will be one-tailed since a small value of X^2 indicates good agreement between Os and Es. Only a large value of X^2 will lead to **H$_0$** being rejected.

Table 6 Percentage points of the χ^2 distribution

The table gives the values of x satisfying $P(X \leqslant x) = p$, where X is a random variable having the χ^2 distribution with ν degrees of freedom.

ν \ p	0.005	0.01	0.025	0.05	0.1	0.9	0.95	0.975	0.99	0.995	ν \ p
1	0.00004	0.0002	0.001	0.004	0.016	2.706	3.841	5.024	6.635	7.879	1
2	0.010	0.020	0.051	0.103	0.211	4.605	5.991	7.378	9.210	10.597	2
3	0.072	0.115	0.216	0.352	0.584	6.251	7.815	9.348	11.345	12.838	3
4	0.207	0.297	0.484	0.711	1.064	7.779	9.488	11.143	13.277	14.860	4

For a 5% significance level the critical value is 5.991. Since $X^2 = 11.6$ and exceeds 5.991, $\mathbf{H_0}$ is rejected and we conclude that the colour is not independent of where the snail was found. This is all that can be concluded from the hypothesis test. However, if the null hypothesis is rejected, an examination of the table will usually suggest some further interpretation which will make the result more informative. In this case the table suggests that snails found on the pavement tend to be lighter coloured than those found in the woodland.

More light snails were observed than were expected on the pavement.

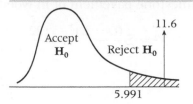

Worked example 7.1

In 1996 Prestbury School entered 45 candidates for A level statistics, while Gorton School entered 34 candidates. The following table summarises the grades obtained.

	A or B	C or D	E	N or U
Prestbury School	8	18	11	8
Gorton School	16	8	5	5

(a) Test at the 5% significance level whether the grades obtained are independent of the school.

(b) Which school has the better results? Explain your answer.

(c) Give two reasons why the school with the better results may not be the better school.   [A]

Solution

(a) $\mathbf{H_0}$ grades obtained are independent of school
$\mathbf{H_1}$ grades obtained not independent of school

The following table shows in, each cell, the observed number on the left-hand side and the expected number (assuming the null hypothesis is true) on the right-hand side. For example the expected number obtaining A or B grades at Prestbury school is $\dfrac{45 \times 24}{79} = 13.67$

	A or B	C or D	E	N or U	Total
Prestbury School	8, 13.67	18, 14.81	11, 9.11	8, 7.41	45
Gorton School	16, 10.33	8, 11.19	5, 6.89	5, 5.59	34
Total	24	26	16	13	79

$$X^2 = \sum \frac{(O - E)^2}{E} = \frac{(8 - 13.67)^2}{13.67} + \frac{(18 - 14.81)^2}{14.81}$$

$$+ \frac{(11 - 9.11)^2}{9.11} + \frac{(8 - 7.41)^2}{7.41}$$

$$+ \frac{(16 - 10.33)^2}{10.33} + \frac{(8 - 11.19)^2}{11.19}$$

$$+ \frac{(5 - 6.89)^2}{6.89} + \frac{(5 - 5.59)^2}{5.59}$$

$$= 8.08$$

> The *E*s have been rounded to 2 dp. Despite this the calculated value of X^2 is correct to 3 sf.

The *O*s are frequencies and the *E*s are all greater than five and so we can compare the calculated value of X^2 with a critical value from the χ^2 distribution.

This is a 2×4 contingency table and so the degrees of freedom are $(2 - 1)(4 - 1) = 3$.

For a 5% significance level the critical value is 7.815.

H₀ is rejected and we conclude that the grades obtained are not independent of the school.

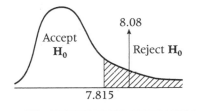

> Don't forget the test is one-tailed.

(b) From the table you can see that Prestbury school got less A or B grades than expected and more of the lower grades than expected. Gorton school got more A or B grades than expected and less of the lower grades than expected. The hypothesis test shows that this is unlikely to have occurred by chance and so you can conclude that Gorton school had the better results.

(c) A level results are only one aspect of a school's worth and cannot on their own be used as a measure of how good a school is. The analysis takes no account of the different intakes of the two schools.

Worked example 7.2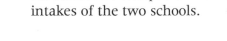

In 1996 Ardwick School entered 130 candidates for GCSE of whom 70% gained five or more passes at grade C or above. The figures for Bramhall School were 145 candidates of whom 60% gained five or more passes at grade C or above, and for Chorlton School were 120 candidates of whom 65% gained five or more passes at grade C or above.

Form these data into a contingency table and test whether the proportion of candidates obtaining five or more passes at grade C or above is independent of the school.

7

Solution

A contingency table must show frequencies, not percentages and the classes must not overlap. Thus for each school it is necessary to calculate the number of candidates who gained five or more passes at grade C and above and the number of candidates who did not do so.

For Ardwick school $\dfrac{70 \times 130}{100} = 91$ candidates gained five or more passes at grade C and above and $130 - 91 = 39$ did not.

Similarly for Bramhall school $\dfrac{60 \times 145}{100} = 87$ did and $145 - 87 = 58$ did not and for Chorlton school $\dfrac{65 \times 120}{100} = 78$ did and $120 - 78 = 42$ did not.

> If the table showed, for each school, the total candidates and the number who gained five or more passes at grade C, it would contain the same information but it would not be a contingency table.

	Ardwick	Bramhall	Chorlton
5 or more passes	91	87	78
<5 passes	39	58	42

H_0 passing five or more GCSEs at grade C is independent of school
H_1 passing five or more GCSEs at grade C is not independent of school

The following table shows in each cell, the observed number on the left-hand side and the expected number (assuming the null hypothesis is true) on the right-hand side. For example, the expected number obtaining five or more GCSE pass grades at Ardwick school is $\dfrac{256 \times 130}{395} = 84.25$.

	Ardwick	Bramhall	Chorlton	Total
5 or more passes	91, 84.25	87, 93.97	78, 77.77	256
<5 passes	39, 45.75	58, 51.03	42, 42.23	139
Total	130	145	120	395

$$X^2 = \Sigma \frac{(O - E)^2}{E} = 3.01$$

The Os are frequencies and the Es are all greater than five and so we can compare the calculated value of X^2 with a critical value from the χ^2 distribution. This is a 2×3 contingency table and so the degrees of freedom are $(2 - 1)(3 - 1) = 2$.

For a 5% significance level the critical value is 5.991. Hence the null hypothesis is accepted and we conclude that there is no convincing evidence to show a difference in the proportions obtaining five or more GCSEs at grade C at the three schools.

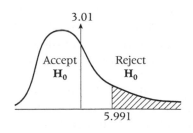

EXERCISE 7A

1 A dairy farmer kept a record of the time of delivery of each calf and the type of assistance the cow needed. Part of the data is summarised below.

	Day	Night
Unattended	42	58
Farmer assisted	63	117
Supervised by a vet	85	35

Investigate, at the 1% significance level, whether there is an association between time of day and type of birth.

2 In a survey on transport, electors from three different areas of a large city were asked whether they would prefer money to be spent on general road improvement or on improving public transport. The replies are shown in the following contingency table:

	Area		
	A	**B**	**C**
Road improvement preferred	22	46	24
Public transport preferred	78	34	36

Test, at the 1% significance level, whether the proportion favouring expenditure on general road improvement is independent of the area.

3 A statistics conference, lasting four days, was held at a university. Lunch was provided and on each day a choice of a vegetarian or a meat dish was offered for the main course. Of those taking lunch, the uptake was as follows:

	Tuesday	**Wednesday**	**Thursday**	**Friday**
Vegetarian	17	24	21	16
Meat	62	42	38	22

Test at the 5% significance level, whether the choice of dish for the main course was independent of the day of the week.

4 A survey into women's attitudes to the way in which women are portrayed in advertising was carried out for a regional television company to provide background information for a discussion programme. A questionnaire was prepared and interviewers approached women in the main shopping areas of Manchester and obtained 567 interviews.

(a) Explain why the women interviewed could not be regarded as a random sample of women living in the Manchester area.

(b) The respondents were classified by age and the following table gives the number of responses to the question 'Do you think that the way women are generally portrayed in advertising is degrading?'

	Under 35 years	35 years and over
yes definitely	85	70
yes	157	126
no	48	11
definitely not	12	3
no opinion or don't know	34	21

Use a χ^2 test, at the 5% significance level, to test whether respondents' replies are independent of age. [A]

5 A private hospital employs a number of visiting surgeons to undertake particular operations. If complications occur during or after the operation the patient has to be transferred to the NHS hospital nearby where the required back-up facilities are available.

A hospital administrator, worried by the effects of this on costs examines the records of three surgeons. Surgeon *A* had six out of her last 47 patients transferred, surgeon *B* four out of his last 72 patients, and surgeon *C* 14 out of his last 41.

(a) Form the data into a 2 × 3 contingency table and test, at the 5% significance level, whether the proportion transferred is independent of the surgeon.

(b) The administrator decides to offer as many operations as possible to surgeon *B*. Explain why and suggest what further information you would need before deciding whether the administrator's decision was based on valid evidence. [A]

6 The following table gives the number of candidates taking AEB Advanced Level Mathematics (Statistics) in June 1984, classified by sex and grade obtained.

	Grade obtained		
	A, B or C	D or E	O or F
Male	529	304	496
Female	398	223	271

In 1984 candidates who just failed to obtain grade E were awarded an O level. Those failing to achieve this were graded F.

(a) Use the χ^2 distribution and a 1% significance level to test whether sex and grade obtained are independent.

(b) Which sex appears to have done better? Explain your answer.

(c) The following table gives the percentage of candidates in the different grades for all the candidates taking Mathematics (Statistics) and for all the candidates taking Mathematics (Pure and Applied) in June 1984.

	Percentage of candidates		
	A, B or C	D or E	O or F
Mathematics (Statistics)	41.5	23.8	34.7
Mathematics (Pure and Applied)	29.2	26.2	44.6

   (i) Explain what further information is needed before a 3×2 table can be formed from which the independence of subject and grade can be tested.

   (ii) When such a table was formed the calculated value of $\sum \dfrac{(O - E)^2}{E}$ was 131.6. Carry out the test using a 0.5% significance level.

   (iii) Discuss, briefly, whether this information indicates that it is easier to get a good grade in 'Statistics' than in 'Pure and Applied'. [A]

7 Analysis of the rate of turnover of employees by a personnel manager produced the following table showing the length of stay of 200 people who left a company for other employment.

	Length of employment (years)		
Grade	0–2	2–5	>5
Managerial	4	11	6
Skilled	32	28	21
Unskilled	25	23	50

This is a 3×3 table. The *E*s are calculated from the sub-totals in exactly the same way as in the previous examples.

Using a 1% level of significance, analyse this information and state fully your conclusions.

7.2 Small expected values

If the expected values are small then, although it is still possible to calculate $\sum \dfrac{(O - E)^2}{E}$, it is no longer valid to obtain a critical value from the χ^2 distribution. This problem can usually be overcome by combining classes together to increase the expected values. There are three main points to bear in mind when considering this procedure:

This is because if *E* is small, a small difference between *O* and *E* can lead to a relatively large value of $\dfrac{(O - E)^2}{E}$.

- It is the expected values, E, which should be reasonably large; small observed values, O, do not cause any problem.

- The classes must be combined in such a way that the data remains a contingency table.

- In order to be able to interpret the conclusions, classes with small Es should be combined with the most similar classes.

> A rule of thumb is that all Es should be greater than five.

> Having decided to combine classes look at the nature of the classes **not** the size of the Es in other classes.

 If it is necessary to combine classes to increase the size of the Es, the most similar classes should be combined.

Worked example 7.3

A university requires all entrants to a science course to study a non-science subject for one year. The non-science subjects available and the number of students of each sex studying them are shown in the table below.

	French	Poetry	Russian	Sculpture
Male	2	8	15	10
Female	10	17	21	37

Use a χ^2 test at the 5% significance level to test whether choice of subject is independent of gender.

Solution

H_0 choice of subject is independent of gender
H_1 choice of subject is not independent of gender

Calculating the expected values in the usual way and writing them on the right-hand side of each cell gives the following table:

	French	Poetry	Russian	Sculpture	Total
Male	2, 3.50	8, 7.29	15, 10.50	10, 13.71	35
Female	10, 8.50	17, 17.71	21, 25.50	37, 33.29	85
Total	12	25	36	47	120

> As there is only one small E and it is only a little less than five this is a borderline case. However it is safer to stick to the rule that all Es should be greater than five.

The expected value for males taking French is less than five. Choose the most similar subject to combine this with. In this case Russian is clearly the appropriate choice as it is the only other language. (We cannot just combine the French and Russian for males as this would leave three cells for males and four cells for females and there would no longer be a contingency table.)

The amended table is as follows:

	Language	Poetry	Sculpture	Total
Male	17, 14.00	8, 7.29	10, 13.71	35
Female	31, 34.00	17, 17.71	37, 33.29	85
Total	48	25	47	120

$$X^2 = \Sigma \frac{(O - E)^2}{E} = 2.42$$

There are two degrees of freedom and for a 5% significance level the critical value is 5.991. Hence the null hypothesis is accepted and we conclude that there is no convincing evidence to show a difference in the choices of males and females.

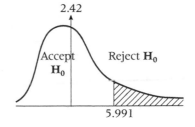

Worked example 7.4

A small supermarket chain has a branch in a city centre and also at an out-of-town shopping centre eight miles away. An investigation into the mode of transport used to visit the stores by a random sample of shoppers yielded the data below.

	Mode of transport			
	Bicycle	**Public transport**	**Private car/taxi**	**Walk**
City centre branch	6	20	36	8
Out-of-town branch	2	9	40	1

(a) (i) Investigate, at the 5% significance level, whether the mode of transport is independent of the branch.

(ii) Describe any differences in the popularity of the different modes of transport used to visit the two branches.

The supermarket chain now extends the investigation and includes data from all of its 14 branches. A 4×14 contingency table is formed and the statistic

$$\Sigma \frac{(O - E)^2}{E}$$

is calculated, correctly, as 56.2 with no grouping of cells being necessary.

(b) Investigate the hypothesis that the mode of transport is independent of the branch:

(i) using a 5% significance level,

(ii) using a 1% significance level.

(c) Compare and explain the conclusions you have reached in **(b) (i)** and **(ii)**. [A]

Solution

(a) **(i)** H_0 mode of transport is independent of branch
H_1 mode of transport is not independent of branch

Calculating the expected values in the usual way and writing them on the right-hand side of each cell gives the following table.

	Mode of transport				
	Bicycle	**Public transport**	**Private car/taxi**	**Walk**	**Total**
City centre	6, 4.59	20, 16.64	36, 43.61	8, 5.16	70
Out-of-town	2, 3.41	9, 12.36	40, 32.39	1, 3.84	52
Total	8	29	76	9	122

The expected numbers going by bicycle to each branch are below five. The most similar form of transport to cycling is walking since both are self-propelled. Fortunately combining cycling and walking also eliminates the problem of the expected number walking to the out-of-town branch being less than five.

The table now becomes

	Mode of transport			
	Bicycle/ Walk	**Public transport**	**Private car/taxi**	**Total**
City centre	14, 9.75	20, 16.64	36, 43.61	70
Out-of-town	3, 7.25	9, 12.36	40, 32.39	52
Total	17	29	76	122

> If the Es had been calculated to greater accuracy (say 7 d.p.) then the value of X^2 would be 9.04 (to 3 s.f.). Small inaccuracies will occur occasionally if the Es are rounded to 2 d.p. but they are generally of no importance.

$$X^2 = \Sigma \frac{(O - E)^2}{E} = 9.05$$

There are two degrees of freedom and for a 5% significance level the critical value is 5.991. Hence the null hypothesis is rejected and we conclude that the method of transport is not independent of branch.

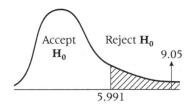

(ii) Examining the table it can be seen that at the city centre branch the observed value for bicycle/walk and for public transport exceeded the expected values whereas the observed value for private car/taxi was less than expected. It therefore appears that private cars or taxis are less likely to be used when visiting the city centre branch than when visiting the out-of-town branch.

(b) (i) The expected values for a 4 × 14 contingency table are calculated in exactly the same way as in any of the tables above. There are $(4 - 1)(14 - 1) = 39$ degrees of freedom. Using a 5% significance level the critical value is 54.572 which is less than 56.2 and so we would conclude that mode of transport is not independent of branch.

> You have been given the value of X^2 as there would not be time in an examination to calculate the Es for such a large table.

(ii) Using a 1% significance level the critical value is 62.428 which is greater than 56.2 and so we would accept that the mode of transport is independent of branch.

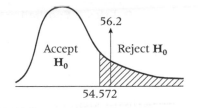

(c) The conclusions in **(b)** mean that if you are prepared to accept a 5% risk of claiming that mode of transport is not independent of branch when in fact it is independent of branch, then you can conclude that mode of transport is not independent of branch. If however you are only prepared to accept a 1% risk, there is insufficient evidence to conclude that mode of transport is not independent of branch.

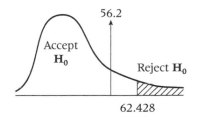

EXERCISE 7B

1 A market researcher is required to interview residents of small villages, aged 18 years and over. She has been allocated a quota of 50 males and 80 females. The age and sex distribution of the her interviewees are summarised below.

Age Group	Male	Female
18–29	3	5
30–39	29	21
40–49	13	40
60 and over	5	14

Investigate, using a 5% significance level, whether there is an association between the sex and age of her interviewees.

2 A manufacturer of decorating materials shows a panel of customers five different colours of paint and asks each one to identify their favourite. The choices classified by sex are shown in the table below.

	White	Pale green	Red	Dark green	Black
Male	8	2	32	9	8
Female	14	5	24	8	10

Test whether choice is independent of gender.

3 The number of communications received, during a particular week, by the editor of a local newspaper are shown in the table below. They have been classified by subject and by whether they were received by letter or by email.

	Politics	Football	Sport (other than football)	Miscellaneous
Letter	20	16	3	27
Email	15	8	2	10

Investigate, using a 1% significance level whether method of communication is associated with subject.

4 A survey is to be carried out among hotel guests to discover what features they regard as important when choosing an hotel. In a pilot study guests were asked to rate a number of features as important or not important. The results for four features are shown below (the number rating each feature is not the same, as not all guests were asked to rate the same features).

	Adequate lighting for reading in bed	Comfortable beds	Courteous staff	Squash courts available
Important	28	34	26	4
Not important	12	17	10	29

(a) Test, at the 5% significance level, whether the proportion of guests rating a feature important is independent of the feature.

(b) Comment on the relative importance of the four features.

(c) Under what circumstances would it be necessary to pool the results from more than one feature in order to carry out a valid test?

(d) If the results for 'adequate lighting for reading in bed' had to be pooled with another feature, which one would you choose and why?

(e) In the final survey 30 features were rated as unimportant, important or very important. The analysis of the resulting contingency table led to a value of 82.4 for $\sum \frac{(O-E)^2}{E}$ (no features were pooled). Test, at the 5% significance level, whether the rating was independent of the feature. [A]

7.3 Yates' correction for 2 × 2 contingency tables

Given the appropriate conditions $X^2 = \Sigma\dfrac{(O - E)^2}{E}$ can be approximated by a χ^2 distribution. In the case of a 2 × 2 contingency table the approximation can be improved by using $\Sigma\dfrac{(|O - E| - 0.5)^2}{E}$ instead of $\Sigma\dfrac{(O - E)^2}{E}$. This is known as Yates' correction.

> The underlying reason for this is that the Os are discrete but the χ^2 distribution is continuous. Hence this is often called Yates' continuity correction.

> For a 2 × 2 table, $\Sigma\dfrac{(|O - E| - 0.5)^2}{E}$ should be calculated. This is known as Yates' correction.

> $|x|$ means the numerical value of x. Thus $|6| = 6$ and $|-3| = 3$.

Worked example 7.5

A university requires all entrants to a science course to study a non-science subject for one year. In the first year of the scheme entrants were given the choice of studying French or Russian. The number of students of each sex choosing each language is shown in the following table:

	French	Russian
Male	39	16
Female	21	14

Use a χ^2 test at the 5% significance level to test whether choice of language is independent of gender.

Solution

H$_0$ Subject chosen is independent of gender
H$_1$ Subject chosen is not independent of gender

Calculating expected values in the usual way and writing them in the right-hand side of the cell gives,

	French	Russian	Total
Male	39, 36.67	16, 18.33	55
Female	21, 23.33	14, 11.67	35
Total	60	30	90

	O	E	$O-E$	$\|O-E\|-0.5$	$\dfrac{(\|O-E\|-0.5)^2}{E}$
Male/French	39	36.67	2.33	1.83	0.091
Male/Russian	16	18.33	−2.33	1.83	0.183
Female/French	21	23.33	−2.33	1.83	0.144
Female/Russian	14	11.67	2.33	1.83	0.287

Be careful to find the modulus of $O - E$ (i.e. $|O - E|$) **before** subtracting 0.5.

$$\sum \frac{(|O-E|-0.5)^2}{E} = 0.705$$

There are $(2-1) \times (2-1) = 1$ degrees of freedom. Critical value for 5% significance level is 3.841.

Accept that choice of subject is independent of gender.

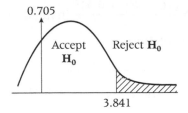

EXERCISE 7C

1 Two groups of patients took part in an experiment in which one group received an anti-allergy drug and the other group received a placebo. The following table summarises the results:

	Drug	Placebo
Allergies exhibited	24	29
Allergies not exhibited	46	21

Investigate, at the 5% significance level, whether the proportion exhibiting allergies is associated with the treatment.
Is the drug effective?
Explain your answer.

2 Of 120 onion seeds of variety A planted in an allotment 28 failed to germinate. Of 45 onion seeds of variety B planted in the allotment four failed to germinate. Form the data into a 2×2 contingency table and test whether the proportion failing to germinate is associated with the variety of seed.

3 Castings made from two different moulds were tested and the results are summarised in the following table:

	Mould 1	Mould 2
Satisfactory	88	165
Defective	12	15

Is the proportion defective independent of mould?

4 As part of a research study into pattern recognition, subjects were asked to examine a picture and see if they could distinguish a word. The picture contained the word 'technology' written backwards and camouflaged by an elaborate pattern. Of 23 librarians who took part 11 succeeded in recognising the word whilst of 19 designers, 13 succeeded. Form the data into a 2 × 2 contingency table and test at the 5% significance level, using Yates' continuity correction, whether an equal proportion of librarians and designers can distinguish the word. [A]

5 The data below refer to the 1996 general election in New Zealand. They show the winning party and the percentage turnout in a sample of constituencies.

Constituency	Winning party	Percentage turnout	Constituency	Winning party	Percentage turnout
Albany	National	87.42	Hunua	National	85.50
Aoraki	Labour	87.95	Hutt South	Labour	87.85
Auckland Central	Labour	88.08	Ham	National	89.22
Banks Peninsula	National	89.71	Kaikoura	National	86.56
Bay of Plenty	National	85.23	Karapiro	National	84.23
Christchurch Cent.	Labour	83.68	Mana	Labour	87.67
Christchurch East	Labour	86.10	Mangere	Labour	79.78
Clutha-Southland	National	84.75	Maungakiekie	National	85.28
Coromandel	National	86.47	Napier	Labour	86.95
Dunedin North	Labour	88.48	Nelson	National	86.88
Dunedin South	Labour	88.87	New Lynn	Labour	85.28
Epsom	National	89.15	North Shore	National	88.34
Hamilton East	National	86.89	Otaki	Labour	88.22
Hamilton West	National	85.09	Owairaka	Labour	86.74

(a) Classify the percentage turnout as 'less than 87' or 'greater than or equal to 87'. Hence draw up a 2 × 2 contingency table suitable for testing the hypothesis that the winning party is independent of the percentage turnout.

(b) Carry out the test using a 5% significance level.

(c) Identify a feature of the constituencies in the sample which suggests that they were not randomly selected. Discuss briefly whether this is likely to affect the validity of the test carried out in **(b)**. [A]

MIXED EXERCISE

1 A well-known picture of The Beatles shows them on a pedestrian crossing outside the Abbey Road recording studios. Substantial numbers of Beatles' fans visit Abbey Road and use the pedestrian crossing.

It was claimed that these fans were causing delays to rush hour traffic. As part of an investigation, people using the crossing were asked whether they were using it in the course of their normal daily lives or because of the Beatles' photograph. The answers and the times of day are summarised in the contingency table below.

	Time of day	
	Rush hour	**Out of rush hour**
Normal daily lives	34	45
Because of Beatles	4	28

(a) Using the 5% significance level, examine whether the reason for using the crossing is associated with the time of day.

(b) Interpret your result in the context of this question.

2 School inspectors in a European country classify some of the lessons they observe as unsatisfactory. Four inspectors worked as a team. Joe Stern observed 138 lessons and classified 13 as unsatisfactory, Janet Grim observed 114 lessons and classified 26 as unsatisfactory, Chris Rough observed 96 lessons and classified 28 as unsatisfactory and Ann de Sade observed 108 lessons and classified 9 as unsatisfactory.

(a) Form the data into a contingency table. Hence investigate, at the 1% significance level, whether the proportion of lessons classified as unsatisfactory is independent of the inspector.

(b) Martin is a teacher whose lesson has been classified as unsatisfactory by Ann de Sade. He claims that the inspectors' judgements are inconsistent and therefore the suggestion that his lesson was unsatisfactory should not be taken seriously. To what extent do your calculations in (a) support Martin's claim? Explain your answer.

(c) In order to investigate the suggestion of inconsistency, the four inspectors all observed the same two lessons and independently awarded marks out of 20. The results were as follows:

	Lesson 1	**Lesson 2**
J. Stern	17	16
J. Grim	12	14
C. Rough	9	12
A. de Sade	15	12

It is suggested that this data should be analysed as a contingency table. Without carrying out any calculations, comment on this suggestion.

3 A college office stocks statistical tables and various items of stationery for salc to students.
The number of items sold by the office and the percentage of these which were statistical tables are shown in the table below. The table gives figures for each of the first five weeks of the autumn term.

Week	1	2	3	4	5
Total number of items sold	216	200	166	105	64
% of items which were statistical tables	11	16	12	17	14

(The percentages have been rounded to the nearest whole number.)

(a) Form the data above into a table suitable for testing whether the proportion of items which were statistical tables is independent of the week.

(b) Carry out this test at the 5% significance level.   [A]

4 A biology student observed snails on a bare limestone pavement and in a nearby woodland. The colour of each snail was classified as light or dark. The following table shows the number of snails observed in each category.

	Light	**Dark**
Pavement	17	10
Woodland	3	15

(a) Use a χ^2 test at the 5% significance level to investigate whether there is an association between the colour of snails and their habitat.

(b) Describe, briefly, the nature of the association, if any, between colour and habitat.

The biology teacher complained that, since fewer than five snails were observed in the Light/Woodland cell, the test carried out in **(a)** was not valid.

(c) Comment on this complaint.

7

The weight, in grams, of a randomly selected snail in each category is recorded in the following table:

	Light	Dark
Pavement	14	12
Woodland	18	21

(d) The geography teacher suggested that a χ^2 test should be applied to these data. Comment on this suggestion. [A]

5 As part of a social survey, one thousand randomly selected school leavers were sent a postal questionnaire in 1996. Completed questionnaires were returned by 712 school leavers. These 712 school leavers were asked to complete a further questionnaire in 1997. The table below shows their response to the 1997 questionnaire classified by their answers to a question on truancy in the 1996 survey.

	Persistent truant	Occasional truant	Never truant
Returned 1997 questionnaire	17	152	295
Failed to return 1997 questionnaire	23	104	121

(a) Use the χ^2 distribution, at the 5% significance level, to test whether returning the 1997 questionnaire is independent of the answer to the 1996 question on truancy.

(b) On the evidence available, state whether 1996 school leavers who played truant are more or less likely to return the 1997 questionnaire. Give a reason for your answer.

(c) A researcher estimated that the proportion of school leavers who played truant persistently in their last year at school was $\dfrac{40}{712}$.

Give two reasons why this might be an underestimate. [A]

6 The following data are from *The British Medical Journal*. The table shows whether or not the subjects suffered from heart disease and how their snoring habits were classified by their partners.

	Never snores	Occasionally snores	Snores nearly every night	Snores every night
Heart disease	24	35	21	30
No heart disease	1355	603	192	224

(a) Use a χ^2 test, at the 5% significance level, to investigate whether frequency of snoring is related to heart disease.

(b) On the evidence above, do heart disease sufferers tend to snore more or snore less than others? Give a reason for your answer.

(c) Do these data show that snoring causes heart disease? Explain your answer briefly.   [A]

7 An incurable illness, which is not life threatening, is usually treated with drugs to alleviate painful symptoms. A number of sufferers agreed to be placed at random into two groups. The members of one group would undergo a new treatment which involves major surgery and the members of the other group would continue with the standard drug treatment. Twelve months later a study of these sufferers produced information on their symptoms which is summarised in the following 2×4 contingency table:

	No change	Slight improvement	Marked improvement	Information unobtainable
New treatment	12	32	46	44
Standard treatment	36	39	12	33

(a) Test at the 5% significance level whether the outcome is independent of the treatment.

(b) Comment on the effectiveness of the new treatment in the light of your answer to (a).

(c) Further analysis of the reasons for information being unobtainable from sufferers showed that of the 44 who underwent the new treatment 19 had died, 10 had refused to cooperate and the rest were untraceable. Of the 33 who continued with the standard treatment three had died, 12 had refused to cooperate and the rest were untraceable. Form these data into a 2×3 contingency table but do not carry out any further calculations. Given that for the contingency table you have formed $\Sigma \dfrac{(O - E)^2}{E} = 10.74$, test, at the 5% significance level, whether the reason for information being unobtainable is independent of the treatment.

(d) Comment on the effectiveness of the new treatment in the light of all the information in this question.   [A]

Key point summary

1 A contingency table shows the frequencies of two   *p140*
   (or more) variables simultaneously.

2 The expected number in any cell of a contingency   *p141*
   table is

$$\frac{(\text{row total}) \times (\text{column total})}{(\text{grand total})}.$$

3 $X^2 = \Sigma \dfrac{(O-E)^2}{E}$ may be approximated by the   *p142*

   χ^2 distribution provided:

   (i) the Os are frequencies,

   (ii) the Es are reasonably large, say >5.

4 An $m \times n$ contingency table has $(m-1)(n-1)$ *p142*
   degrees of freedom.

5 If it is necessary to combine classes to increase the   *p150*
   size of the Es the most similar classes should be
   combined.

6 For a 2×2 table, $\Sigma \dfrac{(|O-E|-0.5)^2}{E}$ should be   *p155*
   calculated. This is known as Yates' correction.

Test yourself	What to review
1 How many degrees of freedom has a 4×3 contingency table?	*Section 7.1*
2 When is Yates' correction applied?	*Section 7.3*
3 Find the appropriate critical value for analysing a 5×4 contingency table, using a 1% significance level.	*Section 7.1*
4 Why are two-sided tests not generally used when analysing contingency tables?	*Section 7.1*
5 A 2×4 contingency contains one cell with an expected value less than five. Why is it incorrect to combine this cell with a neighbouring cell and calculate $X^2 = \Sigma \dfrac{(O-E)^2}{E}$ for the resulting seven cells?	*Section 7.2*
6 Forty males and 50 females choose their favourite colour. A total of 24 choose red. If the results are tabulated in a contingency table, find the expected value for the female/red cell.	*Section 7.2*

7

1 6.

2 When a 2 × 2 contingency table is analysed.

3 26.2.

4 A low value of $\Sigma \dfrac{(O-E)^2}{E}$ indicates very good agreement between observed and expected values and hence is not a reason for rejecting the null hypothesis.

5 The resulting seven cells would no longer form a contingency table.

6 13.3.

CHAPTER 8

Distribution-free methods for single samples and paired comparisons

---

Learning objectives

After studying this chapter, you should be able to:

- understand what is meant by a distribution-free test
- carry out a sign test to test a hypothesis about a population median
- carry out a Wilcoxon signed-rank test to test a hypothesis about a population median
- understand the meaning of experimental error, bias and replication
- carry out a sign test or Wilcoxon signed-rank test on paired data.

8.1 The sign test

This is a very simple hypothesis test which requires no knowledge of the distribution of the population from which the sample is taken.

Many tests require that the population involved is known, or assumed, to be normally distributed.

> Tests which do not require the knowledge or assumption that the data involved is normally distributed are known as **distribution-free** or **non-parametric tests**.

Sign tests involve testing whether the median of a population takes a specific value, or, in some cases, for preferences or differences between populations (see Worked example 8.2 and Section 8.4).

Distribution-free tests can be used when the data is measured on a ratio scale, such as height or weight. They can also be used when only ordinal (rank order) values are available. For the sign test, all that is needed is to be able to decide whether a particular observation is bigger or smaller than another.

A worked example of a sign test is given below. The technical terms involved in hypothesis testing, which you have already met in Chapters 6 and 7, are explained in the context of this example.

> The **sign test** involves allocating a $+$ or $-$ sign to each reading and using a binomial model with $p = \frac{1}{2}$ to determine the critical region.

Worked example 8.1

The lifetime (hours) of a random sample of ten Xtralong lightbulbs taken from a large batch produced in a factory after an expensive machinery overhaul were:

523   556   678   429   558   498   399   515   555   699.

Before the overhaul, the median lifetime of Xtralong bulbs was 500 hours.

Use a sign test, at the 5% **significance level**, to test the hypothesis that the median length of life of Xtralong lightbulbs after the overhaul is greater than 500 hours.

> The **significance level** is the level of overwhelming evidence deemed necessary for the decision to conclude that H_0 is not true. It is the probability of wrongly rejecting a true H_0. The smaller the significance level, the more overwhelming the evidence. See section 6.4.

Solution

H_0 Population median = 500
H_1 Population median > 500

> Note that H_0 must contain = for the sign (or any other) test to be carried out. See section 6.1.

One-tailed test significance level 5%.

If H_0 is true, then

$$P(\text{lifetime} > 500) = P(\text{lifetime} < 500) = \tfrac{1}{2}$$

> If we assume that the median = 500 hours, then half of the population should have a lifetime above 500 and half below 500.

We are interested in whether a lifetime is recorded as greater or less than 500 hours.

We have a sample of 10 lifetimes, so, if H_0 is true then the number of lifetimes greater than 500 hours should follow a B(10, 0.5) distribution.

> The binomial distribution is being used as a model to enable a test to be carried out.

A lifetime greater than 500 hours will be given a + sign and one less than 500 hours a − sign as follows:

523	556	678	429	558	498	399	515	555	699
+	+	+	−	+	−	−	+	+	+

> The use of the + and − is the reason for the test being called the **sign** test.

The cumulative binomial tables will normally provide the probabilities necessary to determine the **critical region**, at the **5% level**, for this test.

> The **critical region** is the range of values which is so unlikely (with probability 0.05 or less since the **significance level** is 5%) to occur that it will lead to the conclusion that H_0 is not true. See section 6.4.

P(0 'more than 500') = 0.0010
P(1 or fewer 'more than 500') = 0.0107
P(2 or fewer 'more than 500') = 0 0547.

> Since the probability of a + is 0.5, we also know, by symmetry, that P(0 'more') = P(10 'more')
> P(0 or 1 'more') = P(9 or 10 'more')
> Hence:
> P(10 'more than 500') = 0.0010
> P(9 or 10 'more than 500') = 0.0107
> P(8, 9 or 10 'more than 500') = 0.0547 and so on.

The number of positive differences possible out of a sample of 10 readings are:

0 1 2 3 4 5 6 7 8 9 10.

In order to identify the **critical region**, we must identify those extreme outcomes which occur 5% of the time **or less**, assuming H_0 true, $p = \tfrac{1}{2}$.

In other words, the probability of these outcomes must be $\leqslant 0.05$.

> Note that the number of negative differences will also follow B(10, 0.5).

8

Since the test is **one-tailed**, only extremely large numbers of positive differences (very small numbers of negative differences) out of the sample of ten need to be considered. The **critical region** will identify the extreme outcomes where high numbers of + signs would occur 5% of the time or less if H_0 is true:

$$0 \quad 1 \quad 2 \quad 3 \quad 4 \quad 5 \quad 6 \quad 7 \quad 8 \quad [9 \quad 10$$
$$\qquad\qquad\qquad\qquad\qquad\quad * \qquad\qquad\qquad P(9, 10) = 0.0107$$

The **critical region** is (9, 10).

The **test statistic** for this test is $7+$ (or $3-$) which is the number of + (or −) signs obtained.
This is labelled * above.

It is clear that the **test statistic** does not lie in the extreme **critical region** and therefore we have no **significant** evidence to doubt that H_0 is true.

Conclusion

No **significant** evidence at **5% level** to doubt H_0 that the median is 500 hours.

Alternative solution

In this example, the test statistic was shown to be 7. Therefore, simply noting that $p = P(\geqslant 7+) = 0.1719$, which is considerably greater than 0.05, for this **one-tailed** test, tells us that the test statistic cannot lie in the **critical region** when a 5% **significance level** is used. There is **no significant** evidence to doubt that H_0 is true.
This method makes the calculations involved in carrying out a sign test more straightforward and is recommended in the exam.
If the relevant p value is **greater** than 0.05, accept H_0.
If the relevant p value is **less** than or equal to 0.05, reject H_0.

A simpler way to carry out a sign test is to examine the probability, p, of obtaining the test statistic or a more extreme value. If p is smaller than or equal to the stated significance level, then H_0 is rejected. Otherwise, H_0 is accepted.

Worked example 8.2

A random sample of 20 children are asked whether they prefer 'Own Brand' (X) or 'Big Name' (Y) breakfast cereal. Their preferences were:

Y X Y Y Y X Y Y Y Y Y Y Y Y Y X Y Y Y Y

Use a sign test, at the 1% significance level, to test the hypothesis that children prefer 'Big Name' breakfast cereal.

The extreme outcome '9 or 10+' has probability **below** 0.05. The outcome '8, 9 or 10+' has probability close to but **above** 0.05. Using the cumulative binomial tables which provide exact probabilities correct to 4 dp, these two facts identify the **critical region** for this test at a 5% **significance level**.

The **test statistic** is a number found from the sample data assuming that H_0 is true.

Note that this is **not** the same as saying that we have **proved H_0** is true.

Conclusions should be in context.

This **alternative** method is probably easier and more informative for the sign test. However, for most other distribution-free tests, calculating the p value is much more complex. Instead you will have to use critical values which can be found directly from tables.

Solution

H_0 The two breakfast cereals are equally desirable
H_1 'Big Name' is preferred

One-tailed test significance level 1%.

If H_0 is true, then

> P('Own Brand' chosen) = P('Big Brand' chosen)
> P(X chosen) = P(Y chosen) = $\frac{1}{2}$

We are interested in whether a preference is recorded for Y, which will be indicated +, or for X, indicated −. For the sample of 20 children, if H_0 is true, then the number of + preferences for Y should follow a B(20, 0.5) distribution.

The number of + preferences for Y and − for X are:

> Y X Y Y Y X Y Y Y Y Y Y Y Y X Y Y Y Y Y
> + − + + + − + + + + + + + + − + + + + +

Again, the cumulative binomial tables will provide the probabilities necessary to determine the **critical region**, at the **1% level**, for this test.

P(4 or fewer 'prefer Y') = P(16 or more 'prefer Y') = 0.0059
P(5 or fewer 'prefer Y') = P(15 or more 'prefer Y') = 0.0207.

The number of + signs possible out of a sample of 20 readings are:

> 0 1 2 3 4 5 6 7 8 9 10 11 12 13 14 15 16 17 18 19 20

In order to identify the **critical region**, we must identify the extreme outcome where high numbers of + occur 1% of the time or less.

The probability of the extremely high number of + must be ⩽0.01.

> 0 1 2 3 4 5 6 7 8 9 10 11 12 13 14 15 [16 17 18 19 20
>   *

>   P(⩾16+) = 0.0059
> but P(⩾15+) = 0.0207

The **critical region** is (16,17,18,19,20+) because P(⩾16+) is **less** than 0.01 but P(⩾15+) **exceeds** 0.01.

The **test statistic** for this test is 17+ (or 3−).
This is labelled * above.

It is clear that the **test statistic** does lie inside the **critical region** and therefore we have overwhelming evidence that H_0 is untrue.

Conclusion

There is **significant** evidence at **1% level** to doubt H_0 that the cereals are equally desirable. We conclude that 'Big Name' is more desirable to children than 'Own Brand' cereal.

H_1 tells us that we expect a lot of Y preferences.

The cumulative binomial tables give cumulative probabilities from P(0) upwards but remember that the distribution is symmetrical since $p = \frac{1}{2}$.
P(0+) = P(20+)
P(⩽1+) = P(⩾19+), etc.

8

As the test is **one-tailed**, we only consider the situation where an extremely high number of + occur for the **critical region**.

Alternative solution

In this example, the test statistic was shown to be 17 and so, noting that $p = P(\geqslant 17+) = 0.0013$ which is considerably **smaller** than 0.01 for this **one-tailed** test, tells us that the test statistic must lie in the **critical region** using a 1% significance level. There is **significant** evidence to reject H_0.

> It is perfectly acceptable to use this alternative method in the exam.

Worked example 8.3

Twenty-one students at a large college undergo a standard test to measure their reaction time to a particular stimulus. The median reaction time in the population generally is believed to be 7.8 seconds.

The reaction times for this sample are:

6.6	3.6	2.1	13.2	5.4	11.6	1.6
7.2	7.8	3.8	6.0	14.2	3.0	15.2
4.7	2.8	7.5	6.9	21.6	6.7	4.3

Use a sign test to test, at the 5% significance level, the hypothesis that the median reaction time is 7.8 seconds.

Solution

H_0 Median reaction time of college students is 7.8 seconds
        Population median = 7.8 seconds
H_1 Median reaction times of students differs from 7.8 seconds
        Population median $\neq$ 7.8 seconds

Two-tailed test significance level 5%.

If H_0 is true, the

$$P(\text{reaction time} > 7.8) = P(\text{reaction time} < 7.8) = \tfrac{1}{2}$$

We only need consider whether a reaction time is recorded as higher than or lower than 7.8 seconds and so the effective sample size involved is 20 since the subject with reaction time 7.8 seconds cannot be included.

A time above 7.8 is given a + sign, one below a − sign.

If H_0 is true, the number of + signs should follow a B(20, 0.5) distribution.

> Providing the proportion of people recording a reaction time of exactly 7.8 seconds is small, the assumption that, if H_0 is true, P(reaction time > 7.8) = P(reaction time < 7.8) = $\tfrac{1}{2}$ will still be reasonable.

6.6−	3.6−	2.1−	13.2+	5.4−	11.6+	1.6−
7.2−	7.8 .	3.8−	6.0−	14.2+	3.0−	15.2+
4.7−	2.8−	7.5−	6.9−	21.6+	6.7−	4.3−

There are five + signs and 15 − signs.

Cumulative binomial tables again provide the probabilities necessary to determine the **critical region**, at the **5% level** for this test.

As this is a **two-tailed** test, those combined extreme outcomes which occur with probability 0.05 or less must be identified.

> Very large and very small numbers of + (or −) signs are both equally relevant for a two-tailed test.

Considering probabilities of extremes,

P(4 or fewer +) = P(16 or more +) = 0.0059
P(5 or fewer +) = P(15 or more +) = 0.0207
P(6 or fewer +) = P(14 or more +) = 0.0577

The number of + signs possible out of a sample of 20 readings are:

0 1 2 3 4 5 6 7 8 9 10 11 12 13 14 15 16 17 18 19 20

Since the test is **two-tailed**, extremely large or small numbers of + signs must both be included and the **critical region** will consider both these extremes combined together.

0 1 2 3 4 5] 6 7 8 9 10 11 12 13 14 [15 16 17 18 19 20+
    *

P($\leqslant$5) = 0.0207	P($\geqslant$15) = 0.0207
P($\leqslant$6) = 0.0577	P($\geqslant$14) = 0.0577

> Each extreme probability will be doubled when **both** extremes are considered for a **two-tailed** test
>
> 0.0207 + 0.0207 = 0.0414, etc.

The **critical region** is (0,1,2,3,4,5,15,16,17,18,19,20+) because P($\geqslant$15) and P($\leqslant$5) = 0.0414 combined together which is **less** than 0.05, but P($\geqslant$14) and P($\leqslant$6) = 0.1154 combined together which **exceeds** 0.05.

The **test statistic** is 15+ (or 5−) which is labelled *.

It is clear that the **test statistic** does lie inside the **critical region** and therefore we have evidence to suggest that H_0 is untrue.

Conclusion

There is **significant** evidence at **5% level** to doubt H_0 that the median reaction time is 7.8 seconds. We therefore conclude that the median reaction time of the population from which this sample was taken differs from 7.8 seconds.

> In this case, because there were a large number of − signs, we think it is very likely that the median reaction time of students is **less** than 7.8 seconds.

8

Alternative solution

The test statistic is 15 and, noting that p = P($\geqslant$15+) + P($\leqslant$5+) = 0.0414 for this **two-tailed** test, which is **smaller** than 0.05, this indicates that there is **significant** evidence to reject H_0.

For a **two-tailed** test, it is sufficient to note that that P($\geqslant$15+) is less than 0.025 as the 5% significance level can be divided into 2.5% or 0.025 in each tail.

> Both extremes together must be considered for two-tailed tests.

EXERCISE 8A

1 A factory produces lengths of rope for use in boatyards. The breaking strength in kilograms for a random sample of 14 lengths of rope were as follows:

        134 136 139 143 136 129 137
        130 138 134 145 141 136 139.

Test the hypothesis that the median breaking strength of all the ropes is 135 kg against the manufacturer's claim that the median breaking strength is greater than 135 kg.
Use the sign test with a 5% level of significance.

2 Twenty-five subjects undergoing a test in a controlled laboratory experiment were recorded as having the following reaction times (seconds) to a particular stimulus:

6.5 3.4 5.6 6.9 7.1   4.9 12.9 7.8 2.4 2.8 15.3 3.7 7.8
2.4 2.8 3.7 4.9 14.0 6.5 22.8 6.9 7.4 3.1 1.9   19.5

Carry out a sign test at the 5% significance level, to test the hypothesis that the median reaction time for the population from which the subjects were drawn is 7.5 seconds.

3 A psychologist carried out an experiment to find how many six letter words 24 randomly chosen students can recall from a list of 20 such words.
The results were:

12	14	15	8	7	10
11	15	17	18	14	15
7	9	10	11	13	13
15	16	12	14	8	12

From previous experiments, it is known that the median number remembered for the population as a whole is 15. Test at the 1% significance level, the hypothesis that students have a lower median recall value than the population as a whole. Use a sign test.

4 A maze is devised and, after many trials on adult participants, it is found that the median length of time to solve the maze is 7.4 seconds.
A group of nine children was then asked to attempt the maze and their times to completion were:

6.1 9.0 8.3 9.4 5.8 8.1 7.6 9.2 10.0 seconds

Use a sign test, at the 5% significance level, to test the hypothesis that children take longer to do the maze than adults.

5 A company devises a trial to determine whether members of the public prefer a sunflower oil based spread to one based on olive oil.

Out of the 30 people involved in this trial, only 10 preferred the olive oil based spread, the rest preferring the sunflower oil spread.

Carry out a sign test, at the 5% level, to determine whether there is a significant preference for the sunflower oil based spread.

6 An ice cream manufacturer is considering introducing a new flavour cornet and asks a panel of 21 tasters for their opinion of the flavour. The scores were given out of a maximum of 100.

Taster	Score	Taster	Score
A	88	L	51
B	94	M	47
C	79	N	33
D	56	O	55
E	67	P	68
F	53	Q	83
G	66	R	62
H	79	S	61
I	83	T	78
J	59	U	90
K	76		

(a) The median flavour score given by all tasters for the original cornet was 52.
Carry out a sign test at the 1% significance level to determine whether the tasters prefer the new flavour to the original flavour.

(b) If you had been asked to test whether the original flavour was preferred to the new flavour, explain why you would not have needed to use the binomial distribution. [A]

7 A motoring correspondent assesses the relative merits of two similarly-priced cars, *A* and *B*, by comparing 32 common features. Her results reveal 21 '+ signs' indicating these features are better on *A*, nine '− signs' indicating these features are better on *B*, and two 'zeros' indicating no difference.

(a) Use binomial tables to investigate at the 5% level of significance, the claim that *A* is the better car.

(b) State **two** assumptions that you have made in reaching your conclusion in **(a)**. [A]

8 It is claimed that adults in the UK visit the cinema on average more than 12 times per year. A sample of 30 adults in the UK contained 23 who had visited the cinema more than 12 times in the last year and 7 who had visited the cinema less than 12 times in the last year.

(a) Use a suitable distribution-free test to investigate the claim at the 1% significance level.

(b) How would your conclusion be affected if you were given the following pieces of information? Answer each part separately.

8

(i) The number of cinema visits does not follow a normal distribution.

(ii) The distribution of the number of cinema visits is not symmetrical.

(iii) The data was gathered by interviewing adults as they left a cinema. [A]

8.2 The Wilcoxon signed-rank test

As with the sign test, the Wilcoxon signed-rank test is a distribution-free or non-parametric test where it is not necessary to know or be able to assume that the data involved is normally distributed. However, the Wilcoxon signed-rank test does need the assumption that the population from which the sample is taken is symmetrically distributed and it can only be used for numerical data.

The Wilcoxon signed-rank test is used to test a hypothesis concerning the mean or the median of a population by considering a single sample.

> Since the population must be symmetrical the mean and the median will be equal.

Unlike the sign test which only considers the signs of the differences between the items in the sample and the suggested value of the population median, the Wilcoxon signed-rank test takes into account the size of these differences and puts them in **rank order**. The Wilcoxon test allows, for example, relatively **few** but very **large** + differences to balance many relatively small − differences. It is therefore to be preferred to the sign test.

The **Wilcoxon signed-rank test** examines the signed differences between each reading and the suggested population mean or median. Rank order values are then assigned to the differences and, for a two-tailed test, the smaller of the totals T^+ or T^- is the test statistic to be compared with the critical value given in Table 10.

The **Wilcoxon signed-rank test** requires a symmetric distribution.

The **Wilcoxon signed-rank test** takes the relative magnitudes of the differences into account and is therefore preferred to the **sign test** provided numerical differences can be obtained.

The easiest way to explain the procedures involved is to work through an example.

Worked example 8.4

The median lifetime of a certain brand of battery is claimed to be 300 hours.

A random sample of 15 of these batteries is taken and their lifetimes recorded:

342	278	302
393	265	289
257	216	312
339	402	249
306	190	178

Use a Wilcoxon signed-rank test, at the 5% significance level, to test whether the claim that the median is 300 hours is justified.

Solution

H_0 Population median = 300
H_1 Population median $\neq$ 300

Two-tailed test significance level 5%.

The first step in carrying out a Wilcoxon test is to find the signed differences between the given data and 300. These are given in the first column of the following table.

Difference	Rank value +	Rank value −
+42	8	
−22		5
+2	1	
+93	12	
−35		6
−11		3
−43		9
−84		11
+12	4	
+39	7	
+102	13	
−51		10
+6	2	
−110		14
−122		15
Totals	$T^+ = 47$	$T^- = 73$

> The differences are found for each figure proceeding along the rows of the original data.

8

The value of the differences are replaced with a **rank order value**. The smallest **absolute difference** is given rank order value 1, the next smallest 2 and so on until the largest difference from the above sample, with 15 valid readings, is given the final rank order value 15.

> The number of differences, $n = 15$.

The rank values are grouped as either + or − differences and the totals T^+ and T^- are calculated, where T^+ is the total of the ranks assigned to positive differences and T^- is the total of the ranks assigned to negative differences.

In this example, $T^+ = 47$ and $T^- = 73$.

In the sign test you were able to calculate the probability of the different outcomes and to deduce the critical region. This is possible but is much more complex for Wilcoxon's test.

Fortunately the critical values have been tabulated. Table 10 provides lower tail critical values of the test statistic T. The critical values are given for one- or two-tailed tests and for several significance levels.

As Table 10 provides **lower** tail critical values, the procedure for carrying out a two-tailed test is to compare T, the smaller of T^+ and T^-, with the value in Table 10.

The relevant critical value for a two-tailed test with n 5 15 readings, at the 5% level of significance, is 25.

Therefore, if T < 25, there is significant evidence of a very extreme result which would lead to H0 being rejected. For this test T is greater than 25.

Conclusion

T is **greater** than the critical value so there is no significant evidence at the 5% level to reject $\mathbf{H_0}$ that the median lifetime of batteries is 300 hours

> A check can be made on the totals since $T^+ + T^- = \frac{1}{2}n(n+1)$. In this case $47 + 73 = 120$ and $\frac{1}{2}n(n+1) = \frac{1}{2}(15)(16) = 120$.

> It is possible, but unnecessary, to derive the upper tail critical values from Table 10. If T is above the critical value in Table 10, then both T^+ and T^- will lie between the upper and lower critical values.

> As you have seen in the sign test, a significance level of exactly, say, 5% cannot usually be obtained. The value shown in Table 10 gives a value as close as possible to that tabulated. Thus the significance level of this critical value may be a little higher or a little lower than 5%. This has no effect on the way the tables are used.

Worked example 8.5

Eleven job applicants are randomly chosen from a large group and asked to attend an interview during which each applicant takes an aptitude test to identify which would be best suited to the job available. The mean score on this test nationally is known to be 64.

The scores of the most recent applicants were:

    56 57 63 64 62 65 56 65 69 60 61

Test the hypothesis, using a 5% significance level, that this group of applicants have a lower aptitude than that found nationally.

Solution

$\mathbf{H_0}$ Mean score in group $= 64$
$\mathbf{H_1}$ Mean score in group < 64

One-tailed test significance level 5%.

Difference	Rank value +	Rank value −
−8		9.5
−7		8
−1		2
0	discard	
−2		4
+1	2	
−8		9.5
+1	2	
+5	7	
−4		6
−3		5
Totals	$T^+ = 11$	$T^- = 44$

> Differences equal to zero must be discarded. This will affect the calculation of the critical values. However, provided only a small proportion of observations are discarded, Table 10 can still be used to find critical values.

In this example, some of the absolute differences are the same. The individual rank values 1, 2, 3, …, 10 cannot be assigned in the usual way and the rule is that to each of the equal differences, the **average** of the ranks they would normally have received is assigned.

The absolute difference 1 occurs three times and those three places should have been allocated the ranks 1, 2 and 3.

The average is $\dfrac{1 + 2 + 3}{3} = 2$ so all receive rank 2.

The absolute difference 8 occurs twice and those 2 places should have been allocated the ranks 9 and 10.
The average of these two ranks is 9.5 so each receives rank 9.5.

In this example, the alternative hypothesis, H_1 is mean <64.
If H_1 is true, we would expect T^- to be large and T^+ to be small. As the lower tail is given in Table 10, the procedure is to compare T^+ with the critical value in Table 10 to see if it is small enough for us to conclude that H_0 can be rejected and H_1 accepted.

Table 10 gives the **one-tailed** critical value at 5% level as 11.

Therefore if $T^+ \leqslant 11$, there is significant evidence of an extreme result which would lead to H_0 being rejected. For this one-tailed test the test statistic is equal to T^+. That is $T = 11$.

> The **critical region** includes the value given in the table.

Conclusion

T is **equal** to the critical value and therefore there is significant evidence at the 5% level to reject H_0 and conclude that the mean score for the group of applicants is lower than 64.

EXERCISE 8B

1 For the data given in Question **1** of Exercise 8A, test the hypothesis again, this time using the Wilcoxon signed-rank test at the 5% level.

2 For the data given in Question **4** of Exercise 8A, test the hypothesis again, this time using the Wilcoxon signed-rank test at a 5% significance level.

3 The birth rate in an African state is believed to have a mean value of 51 births per year per 1000 population.
After an intensive education programme in one particular area of this state, birth rates in 12 large settlements in this area are recorded as:

> 47.5 48.8 47.8 50.1 49.0 52.0
> 46.0 50.3 42.5 47.0 43.6 43.8

Test whether the mean birth rate seems to have declined in this area using a Wilcoxon signed-rank test at the 1% level of significance.
(Remember that for a Wilcoxon test to be carried out, the population must be symmetrical and hence the mean and median are the same. See Section 8.2.)

4 A Spanish teacher sets her A level students a set of words to learn. This vocabulary list contains 30 words and the median number correct in previous years was 23.
The number of words correctly remembered by a sample of 11 students this year are:

> 24 29 27 15 23 30 25 28 21 26 17

Use a Wilcoxon signed-rank test at the 5% level of significance to determine whether there is evidence that the median score in the vocabulary test has changed this year.

5 The following data relates to the amount of money, to the nearest $, spent by a random sample of 16 visitors to a theme park in the USA:

> 84 72 98 108   55 115 68 102
> 78 89 77 105 112   85 69 108

Using the Wilcoxon signed-rank test, test the hypothesis that the median amount spent, per visitor, is $95. Use the 5% significance level.

6 The values below are the scores obtained by a batsman in a random sample of 20 innings in one-day cricket matches.

> 26   0   0   0 103
> 28   16   8   14   0
> 18   47   0   2   0
> 52   25 128   26   84.

The batsman's median score for all innings in four-day cricket matches is 30.

(a) Explain why the Wilcoxon signed-rank test is unsuitable for use with this data.

(b) Using a sign test at the 5% level of significance, investigate the claim that the batsman's median score in one-day cricket matches is less than that in four-day matches. [A]

7   A group of adult males attending a cardiac clinic are assessed on various health factors and are assigned a relative risk value indicating their likelihood of suffering a major coronary event.
The relative risk values for this group were:

    1.6   3.2   1.7   2.7   1.9   4.3   1.6   2.6
    1.1   1.8   1.4   1.9   1.3   1.6   2.3

The median relative risk value for all adult males is 1.5.

(a) Investigate, using a sign test with a 5% significance level, whether there is any evidence to suggest that the adult males attending this cardiac clinic have a median relative risk value greater than 1.5.

The consultant who runs this cardiac clinic is also carrying out research into the blood pressure level and occupation of adult males. The mean blood pressure level for all adult males is 74. A randomly selected sample of male teachers, who were all working in large comprehensive schools, have their blood pressures measured.

The results were:

    92   89   106   84   74   90   71   98   87

(b) Carry out a Wilcoxon signed-rank test, with a 1% significance level, to investigate whether this sample indicates that male teachers have a mean blood pressure higher than 74.

(c) Give two reasons why conclusions regarding the blood pressure levels of **all** teachers cannot be drawn by referring to your test results in **(b)**.   [A]

8   The median life of a make of candle is 270 minutes.
A different make of candle is claimed to have a median life longer than 200 minutes.
To test this, 20 of the new candles are lit and after 200 minutes it is observed six have burnt out but the remainder are still burning.

(a) Use an appropriate distribution-free method to investigate the claim that the median life exceeds 200 minutes. Use a 5% significance level.
The six candles which had burnt out lasted

    162   179   183   184   189   195 minutes.

The next candles to burn out lasted

210   215   225   234 and 239 minutes.

The remaining nine candles were still burning after 240 minutes.

(b) Use Wilcoxon's signed-rank test, at the 1% significance level, to investigate the claim that the median life of the new candles exceeds 200 minutes.

(c) State two assumptions it was necessary to make in order to carry out the test in **(b)**.

(d) Explain why it would not have been possible to apply Wilcoxon's signed-rank test when the sample of candles had been burning for 212 minutes.   [A]

9   The external diameters (measured in units of 0.01 mm above a nominal value) of a random sample of piston rings from a large consignment were:

11   9   32   18   29   1   21   19   6   3.

(a) Use Wilcoxon's signed rank test, at the 5% significance level, to investigate the claim that the median external diameter is 20.

It was later discovered that an error had been made in zeroing the measuring device and that all the measurements in the sample should be increased by 12.

(b) Repeat **(a)** using the correct measurements.

A technician carried out the tests in **(a)** and **(b)** and in each case accepted the null hypothesis. She suspected an error in her calculations because, although the sample median had increased by 12 in **(b)** compared to **(a)**, she had accepted the same conclusion about the median in both cases.

(c) In the context of this example explain the meaning of
  (i)   null hypothesis,
  (ii) Type I error,
  (iii) Type II error.

Hence, explain to the technician why there is no reason for her to conclude that there was an error in her calculations.   [A]

8.3 Experimental design

Variability of experimental results is a fact of life and something we all expect. You may well, when at school, have been given a seed from a sunflower to plant. Everybody had similar sized seeds from the same flower and you all planted them in identical pots supplied by your teacher. They had the same conditions to grow and yet nobody would expect them all to grow to exactly the same height. Indeed, the fun of this trial was often to see whose seed grew into the tallest plant. Such variability of results is called **experimental error** which does not mean that a mistake has been made but rather that there are always other factors affecting results. In this case, some seeds were probably 'stronger' than others, some seeds were slightly over- or under-watered, etc. and some pots were at the sunny end of the windowsill, others were not. **Experimental error** should be minimised by keeping factors which are not being investigated as constant as possible and by careful experimental design.

What factors could be kept constant in an experiment to compare the petrol consumption of two different makes of car?
There are many factors to consider, such as:
same driver used, same route taken, same time of day for trials, same brand of fuel used, same weather conditions.

It is desirable to take **repeated observations** or **replicates** under apparently identical conditions in order to estimate the **experimental error**. In the previous example, if the fuel consumption of a make of car is estimated several times these are **replicates**. If these estimates are made by the same driver, taking the same route, using the same fuel, at the same time of day in the same weather conditions (as far as is feasible), then the **experimental error** will be reduced.

> The purpose of having **replicates** is to estimate the magnitude of the experimental error because the only possible differences would be due to experimental error not due to the factors being investigated.

One of the simplest experimental designs which is often used to reduce **experimental error** is to plan **paired comparisons**. For example, if the response times of school pupils to a particular stimulus are believed to be faster before 10 a.m. than after 4 p.m., one group of 30 pupils could be tested at 9.30 a.m. and another group at 4.30 p.m. However, the fact that pupils are known to vary enormously in their response times, means that it becomes difficult to determine if differences between the groups really are due to the time of day or due to the inbuilt differences between the two groups. A far better design would be to use the same 30 pupils for both tests and then any difference could be attributed to the time of day since no variability between the pupils exists: they are the same pupils!

> You will use two examples of distribution-free tests for analysing paired comparisons later in this chapter: the sign and the Wilcoxon signed-rank test on paired data.

8

Similarly, to compare weight loss due to two different slimming diets, an ideal design would be to secure the cooperation of several pairs of identical twins of similar weights. One twin of each pair would follow one diet and the other twin the second diet. Then, any **experimental error** due to physiological differences in the people following the two diets would be minimised.

> **Experimental design** is used to eliminate **bias** and reduce **experimental error** in data collection.

8.4 The paired sample sign test

The sign test which was introduced in section 8.1 can easily be adapted to test for a difference between paired samples. In the slimming example mentioned above, the aim is to establish which (if either) diet leads to the greater weight loss on average.

> If a paired test is not possible then a test on two **separate** samples of data is the only option. Chapter 9 introduces the Mann–Whitney U test which is the non-parametric test used in this situation.

In order to carry out the sign test, this has to be formulated more precisely. The H_0 used this time is that the differences between the pairs come from a population which has median equal to zero. Under this H_0, the second reading in each pair is equally likely to be greater than or smaller than the first. This is exactly the same situation as in the single sample sign test seen in section 8.1. An example will explain how this works.

> A **paired comparison** is a simple experimental design which can be used to reduce experimental error when two treatments are being compared.

Worked example 8.6

To measure the effectiveness of a drug for asthmatic relief, 12 subjects, all susceptible to asthma, were each administered the drug after one asthma attack and the placebo after a separate asthma attack. One hour after the attack an asthmatic index was obtained on each subject with the following results:

Subject	1	2	3	4	5	6	7	8	9	10	11	12
Drug	28	31	17	18	31	12	33	24	18	25	19	17
Placebo	32	33	23	26	34	17	30	24	19	23	21	24

Making no assumptions regarding the distribution of these data, investigate the claim that the drug significantly reduces the asthmatic index using a sign test at the 5% significance level.

> The use of the word **reduces** indicates that this is a one-tailed test.

Solution

H_0 Population median difference $= 0$
H_1 Population median difference > 0

One-tailed test significance level 5%.

If H_0 is true, then, for the index concerned

$$P(\text{placebo} < \text{drug measure}) = P(\text{placebo} > \text{drug measure})$$
$$P(\text{difference} +) = P(\text{difference} -) = \tfrac{1}{2}$$

The differences are given below.
They are all placebo index–drug index

$$+ \;\; + \;\; + \;\; + \;\; + \;\; + \;\; - \;\; . \;\; + \;\; - \;\; + \;\; +$$

There are 11 valid signs as subject 8 scored 24 both times. As before, the cumulative binomial tables will provide the probabilities necessary to determine the critical region, at the 5% level, for this test.
The model used in this case is B(11, 0.5).

The possible number of + signs for this model are:

$$0 \quad 1 \quad 2 \quad 3 \quad 4 \quad 5 \quad 6 \quad 7 \quad 8 \quad [9 \quad 10 \quad 11+$$

> Where difference is taken as placebo − drug measure.

> It does not matter whether you find differences as placebo − drug, as found here, or as drug − placebo **but** you must be consistent throughout the solution.

> Remember that the sign test requires a + or − sign so a zero difference cannot be included.

From the tables:
P(11+) = 0.0005
P(10 or 11+) = 0.0059
P(9, 10 or 11+) = 0.0327
The **test statistic** is the number of + signs obtained, in this case 9 as indicated * above.
Considering the **test statistic**, you can see that P($\geq$9+) is 0.0327 which is less than 0.05 and so **H₀** is rejected.

Also from the tables, P(8, 9, 10 or 11+) = 0.1133. This identifies the critical region as (9, 10, 11+) since P(9, 10 or 11+) is **less** than 0.05, but P(8, 9, 10 or 11+) is **too big** at 5% level. The **test statistic** 9 lies inside the critical region and so we reject **H₀**.

Conclusion

The test statistic lies in the critical region at the 5% level and so **H₀** is rejected and we conclude that there is significant evidence that the drug does reduce the index.

The sign test can also be used on non-numerical data as Worked example 8.2 illustrates.

Worked example 8.7

Fifteen girls were each given an oral examination and a written examination in French. Their grades (highest = A, lowest = F) in the two examinations were as follows.

Girl	1	2	3	4	5	6	7	8	9	10	11	12	13	14	15
Oral exam	A	B	C	D	F	C	B	E	E	C	D	C	E	C	B
Written exam	B	D	D	C	E	D	C	D	C	D	E	E	F	D	C

Using the sign test, investigate the hypothesis that one examination produces significantly different grades from the other, using a 5% level of significance.

Solution

H₀ Population median difference = 0
H₁ Population median difference $\neq$ 0
Two-tailed test significance level 5%.

If **H₀** is true, then, for the grades,

$$P(\text{oral} < \text{written}) = P(\text{oral} > \text{written})$$
$$P(\text{difference} +) = P(\text{difference} -) = \tfrac{1}{2}$$

The differences, oral − written grade, are given below.

+ + + − − + + − − + + + + + +

The model used in this case is B(15, 0.5).
The possible number of + signs for this model are:

0, 1, 2, 3,] 4, 5, 6, 7, 8, 9, 10, 11, [12, 13, 14, 15+
                                                  *

The **test statistic** is the number of + signs obtained, in this case 11 as indicated * above.
Noting that P($\geq$11+) together with P($\leq$4+), for this two-tailed test, is 0.1184 which is **greater** than 0.05 leads to the conclusion that there is no significance evidence to reject **H₀**.

From the tables:
P(0, 1, 2, 3+) = 0.0176
= P(12, 13, 14, 15+)
This identifies the critical region, considering both extremes together for this **two-tailed** test, as:
(0, 1, 2, 3, 12, 13, 14, 15+)
since the **combined** probability for this event is
0.0176 + 0.0176 = 0.0352 which is less than 0.05.
However, P(0, 1, 2, 3, 4+)
= 0.0592 = P(11, 12, 13, 14 or 15+) which combined gives a total probability of 0.1184 which is **too big** at the 5% level.

8

Conclusion

The test statistic does not lie in the critical region so there is no significant evidence to doubt **H₀**. No significant difference in grades found.

> The test statistic 11 does **not** lie in the critical region and so you can accept **H₀**.

EXERCISE 8C

1 Ten athletes ran a fixed 200 m distance on successive days, firstly on a synthetic athletic track and then on a conventional cinder track. The decision whether each athlete ran on cinder first or synthetic first was made at random. The results in seconds were as follows:

Athlete	1	2	3	4	5	6	7	8	9	10
Synthetic	26.5	25.8	27.2	28.1	25.6	25.5	28.8	27.1	24.1	26.6
Cinder	26.6	26.1	27.4	28.0	25.8	26.6	29.1	27.0	24.8	26.8

Carry out a sign test, at the 5% significance level, to determine whether the nature of the surface influences athletes' performance in the 200 m.

2 Ten psychology students carried out an experiment. They wished to test whether the ability to perform a simple control task is influenced by the presence of an audience.

Each student carried out the task on their own first and measured the time taken. Then, each student performed the same task again in front of an audience. The time results (seconds) were:

Student	A	B	C	D	E	F	G	H	I	J
Alone	45.4	48.2	47.5	49.1	54.3	45.5	58.2	47.1	54.3	46.8
Audience	46.7	51.2	47.8	48.0	55.8	46.6	59.1	47.0	54.8	49.6

(a) Explain why this is known as a paired test.

(b) Explain how experimental error is reduced by using paired data.

(c) Carry out a sign test at the 5% significance level to determine whether students take longer to perform this task when an audience is present.

(d) What problem arises in interpreting the results of this experiment? How could the design have been improved in order to avoid this problem? [A]

3 Pairs of twins, where each twin suffers from moderate eczema, are recruited for the trial of a new skin preparation. The trial is a double blind trial in which the twin selected at random to be in the control group is given a placebo. The percentage improvement after 4 weeks of treatment was assessed with the following results:

Twin	1	2	3	4	5	6	7	8
Placebo	16	10	16	22	22	24	24	11
New prep	21	16	20	25	20	28	26	15

Carry out a sign test, at the 5% significance level, to determine whether the new preparation appears to result in a twin having a higher percentage improvement of their eczema.
[A]

4 Identical programs were run on two different makes of personal computers and the load times (seconds) on each machine, for each program, were noted.

Carry out a distribution-free test on this paired data to determine whether there is any evidence of a difference in load times between the two personal computers. Use the sign test at the 5% level.
[A]

Program	1	2	3	4	5	6	7	8	9	10	11	12
PC A	37	77	49	26	23	16	15	11	45	25	9	55
PC B	30	66	47	22	20	14	17	13	43	31	7	41

5 On the 2nd July 1980 the incoming mail in each of 12 selected towns was randomly divided into two similar lots prior to sorting. In each town one lot was then sorted by the traditional sorting method, the other by a new Electronic Post Code Sensor Device (EPCSD). The times taken, in hours, to complete these jobs are recorded below.

Town	A	B	C	D	E	F	G	H	I	J	K	L
Hand sort time	4.3	4.1	5.6	4.0	5.9	4.9	4.3	5.4	5.6	5.2	6.1	4.7
EPCSD sort time	3.7	5.3	4.5	3.1	4.8	5.0	3.5	4.9	4.6	4.1	5.7	3.5

Use the sign test and a 5% level of significance to test the null hypothesis of no difference in the times against the alternative hypothesis that the EPCSD method is quicker.

What further information would you require before making a decision whether or not to change over to the new EPCSD system?
[A]

8.5 The paired sample Wilcoxon signed-rank test

As you can see from Worked example 8.6, the procedure involved in carrying out a paired sign test is exactly the same as that used for a one sample sign test except that the differences between pairs are used. In much the same way, you already know the procedure to follow for the Wilcoxon signed-rank test on paired data. The following example will explain further.

> The **sign test** and the **Wilcoxon signed-rank test** are **distribution-free tests** which can be used to test for differences between **paired** data.

Worked example 8.8

In a comparison of two computerised methods, A and B, for measuring physical fitness, a random sample of eight people was assessed by both methods. Their scores (maximum 20) were recorded as follows.

Subject	1	2	3	4	5	6	7	8
Method A	11.2	8.6	6.5	17.3	14.3	10.7	9.8	13.3
Method B	10.4	12.1	9.1	15.6	16.7	10.7	12.8	15.5

Use a Wilcoxon signed-rank test, at the 5% significance level, to test whether the claim that Method A gives a lower measure of fitness than Method B is justified.

> Remember that the Wilcoxon test is to be preferred to the sign test, as long as differences can be measured, because it takes into account the relative magnitudes, not just the signs, of the differences. Also, remember that the Wilcoxon signed-rank test requires that the population is symmetrical.

Solution

H_0 Population mean difference $= 0$
H_1 Population mean difference $(A - B) < 0$

One-tailed test significance level 5%.

The first step in carrying out a Wilcoxon test on paired data is to find the signed differences between the pairs.

Difference	Rank value	
(A − B)	+	−
+0.8	1	
−3.5		7
−2.6		5
+1.7	2	
−2.4		4
0	Discard	
−3.0		6
−2.2		3
Totals	$T^+ = 3$	$T^- = 25$

> Since we must assume that the differences are symmetrical in order to carry out the Wilcoxon test, the mean and the median are the same. H_0 and H_1 could refer to the mean or to the median.

Any differences equal to zero must be discarded.

As in the previous examples of the Wilcoxon test, the **absolute value** of the remaining differences are replaced with a **rank order value** and these rank values are grouped as either $+$ or $-$ differences. The rank values and their totals T^+ and T^- are given in the previous table.

Since H_1 is that Method A gives a **lower** measure of fitness than Method B, the null hypothesis H_0 will be rejected if T^+ is small. As Table 10 gives the lower critical tail values, it is always advisable to consider which of T^+ and T^- is expected to be small for a one-tail test and then compare this value with the relevant critical value.

The relevant critical value from Table 10 for a one-tailed test with $n = 7$ readings, at the 5% level of significance, is 4.

Therefore, if $T^+ \leqslant 4$, there is significant evidence of a very extreme result which would lead to H_0 being rejected. For this test $T^+ = 3$.

> You can check your totals
> $T^+ + T^- = \frac{1}{2}n(n + 1)$ always
> $3 + 25 = 28 = \frac{1}{2}7(7 + 1)$

Conclusion

T^+ is less than the critical value so we reject H_0 at the 5% significance level and conclude that there is significant evidence to suggest Method A does give a lower average measure of fitness.

Worked example 8.9

An athletics coach wishes to test the value to athletes of an intensive period of weight training and so selects twelve 400-metre runners from the region and records their times, in seconds, to complete this distance. They then undergo the programme of weight training and have their times, in seconds, for 400 metres measured again. The table below summarises the results.

Athlete	A	B	C	D	E	F	G	H	I	J	K	L
Before	51.0	49.8	49.5	50.1	51.6	48.9	52.4	50.6	53.1	48.6	52.9	53.4
After	50.6	50.4	48.9	49.1	51.6	47.6	53.5	49.9	51.0	48.5	50.6	51.7

Use the Wilcoxon signed-rank test at 5% significance level to investigate the hypothesis that the training programme will significantly improve athletes' times for the 400 metres.

Solution

H_0 Population mean difference $= 0$
H_1 Population mean difference (before $-$ after) > 0

One-tailed test significance level 5%.

> The word *improve* indicates a one-tailed test.

Difference	Rank value	
(before − after)	+	−
+0.4	2	
−0.6		3.5
+0.6	3.5	
+1.0	6	
0	Discard	
+1.3	8	
−1.1		7
+0.7	5	
+2.1	10	
+0.1	1	
+2.3	11	
+1.7	9	
Totals	$T^+ = 55.5$	$T^- = 10.5$

Since H_1 is that the training programme will reduce the times, we will reject H_0 and accept H_1 if T^- is small.

The relevant critical value from Table 10 for a one-tailed test with $n = 11$ readings, at the 5% level of significance, is 14. Therefore, if $T^- \leqslant 14$, there is significant evidence of a very extreme result which would lead to H_0 being rejected. For this test T^- is 10.5, clearly **less** than 14.

> Again, check the T^+ and T^- with
> $$T^+ + T^- = \tfrac{1}{2}n(n + 1)$$
> $$55.5 + 10.5 = 66 = \tfrac{1}{2}11(11 + 1)$$

Conclusion

T is less than the critical value so there is significant evidence at the 5% level to reject H_0 and conclude that there is significant evidence to suggest the mean time before is **greater** than the mean time after training. The training programme does appear to significantly improve athletes' times.

Worked example 8.10

As part of her research into the behaviour of the human memory, a psychologist asked 15 schoolgirls to talk for 5 minutes on 'my day at school'. Each girl was then asked to record how many times she thought that she had used the word 'nice' during this period. The table below gives their replies together with the true values.

Girl	A	B	C	D	E	F	G	H	I	J	K	L	M	N	O
True value	12	20	1	8	0	12	12	17	6	5	24	23	10	18	16
Recorded value	9	21	3	14	4	12	16	14	5	9	20	16	11	17	19

Use Wilcoxon's test to investigate whether schoolgirls tend to underestimate or overestimate the frequency with which they use a particular word in a verbal description. Use a 5% significance level.

Solution

H_0 Population mean difference $= 0$
H_1 Population mean difference (true $-$ recorded) $\neq 0$

Two-tailed test significance level 5%.

Difference (true − recorded)	Rank value +	−
+3	7	
−1		2.5
−2		5
−6		13
−4		10.5
0	Discard	
−4		10.5
+3	7	
+1	2.5	
−4		10.5
+4	10.5	
+7	14	
−1		2.5
+1	2.5	
−3		7
Totals	$T^+ = 43.5$	$T^- = 61.5$

The test involved in this example is two-tailed and, since Table 10 provides lower tail critical values, it is only necessary to compare the smaller of T^+ and T^- with the tabulated value.

In this case the test statistic $T = 43.5$ and the two-tailed critical value for a sample of 14 valid differences is 21.

T is above the critical value and so we accept H_0.

Conclusion

There is no significant evidence to doubt H_0 and so we conclude that there is no evidence that girls tend either to underestimate or to overestimate the frequency with which they use the word 'nice'.

> Again, check the T^+ and T^- with
> $T^+ + T^- = \frac{1}{2}n(n + 1)$
> $43.5 + 61.5 = 105 = \frac{1}{2}14(14 + 1)$

> Remember from Section 8.2, that it is not necessary to obtain the upper tail critical value.

8

> If **T** is above the critical value then both T^+ and T^- will lie between the upper and lower critical values.

MIXED EXERCISE

1 For the data given in Exercise 8C Question **1**:

   (a) Comment briefly on the experimental design used.

   (b) Carry out a Wilcoxon signed-rank test, at the 5% significance level, to determine whether the nature of the surface influences athletes' performance in the 200 m.

ee methods for single samples and paired comparisons

it a Wilcoxon signed-rank test on the data given in
8C Question **2** to investigate whether the median
time taken by students to perform the task is greater when an
audience is present. Use a 5% significance level.

3 Carry out a Wilcoxon signed-rank test on the data given in
Exercise 8C Question **3** to investigate whether the new
preparation results in a twin having a higher mean percentage
improvement of their eczema. Use a 1% significance level.

4 Identical programs were run on two different makes of
personal computers and the load times (seconds) on each
machine, for each program, were noted.

Program	1	2	3	4	5	6	7	8	9	10	11	12
	10	11	3	7	5.5	3	5	1	3	9	5.5	12
PC A	37	77	49	26	23	16	12	12	45	25	10	55
PC B	30	66	47	22	20	14	17	13	43	31	7	41
	+7	+11	+2	+4	+3	+2	+5	-1	+2	-6	+3	+14

Use the Wilcoxon signed-rank test, at the 5% significance level,
to determine whether there is any evidence of a difference in
mean load times between the two personal computers.

5 The blood clotting times for eight people were measured
before and after they had consumed a fixed amount of
alcohol. The times (seconds) are given below:

Person	1	2	3	4	5	6	7	8
Before	124	167	129	117	146	16	119	149
After	126	117	134	127	126	128	114	99

(a) Comment on the use of a paired design for this experiment.

(b) Test, at the 5% significance level, the hypothesis that the
consumption of alcohol has no effect on the median
clotting time of blood. Use a Wilcoxon signed-rank test.

6 The Ministry of Defence is considering which of two shoe
leathers it should adopt for its new Army boot. They are
particularly interested in how boots made from these leathers
wear and so 15 soldiers are selected at random and each
soldier wears one boot of each type. After six months the wear,
in millimetres, for each boot is recorded as follows.

Soldier	1	2	3	4	5	6	7	8	9	10	11	12	13	14	15
	+0.7	-0.6	-0.5	-1.2	-0.3	-0.6		-0.4	-0.7	0.1	-1	-0.8	0.9	0.2	
Leather A	5.4	2.6	4.3	1.1	3.3	6.6	4.4	3.5	1.2	1.3	4.8	1.2	2.8	2.0	6.1
Leather B	4.7	3.2	3.8	2.3	3.6	7.2	4.4	3.9	1.9	1.2	5.8	2.0	3.7	1.8	6.1
	8.9	6.55	13	3	6.5			4	8.5	1	12	10	11	2	

Use the Wilcoxon signed-rank test to investigate the
hypothesis that the wear in the two leathers is the same.
Use a 5% significance level

7 Trace metals in drinking water affect the flavour of the water and high concentrations can pose a health hazard. The following table shows the zinc concentrations, in milligrams per 1000 litres, of water on the surface and on the river bed at each of 12 locations on a river.

Location	1	2	3	4	5	6	7	8	9	10	11	12
Surface	387	515	721	341	689	599	743	541	717	523	524	445
Bed	435	532	817	366	827	735	812	669	808	622	476	387

Using a Wilcoxon signed-rank test, examine the claim that zinc concentration of water in this river is higher on the river bed than on the surface. Use a 1% significance level

8 A random sample of 11 adults, who had eaten breakfast at 8 a.m., had their pulse rates measured at 11 a.m., and then again at 7 p.m. immediately after they eaten their evening meal. The results were:

Person	A	B	C	D	E	F	G	H	J	K	L
Pulse 11 a.m.	62	75	87	80	89	81	84	82	75	59	68
Pulse 7 p.m.	60	69	83	79	87	76	75	84	75	58	69

Test whether there is any significant difference, at the 5% level, between pulse rates mid-morning and pulse rates immediately after an evening meal is eaten.

(a) Carry this test out:

    (i) using a sign test,

    (ii) using a Wilcoxon signed-rank test.

(b) Compare the results from these two distribution-free tests. Why might the Wilcoxon test be preferred to the sign test? [A]

9 Jim, a market trader, decided to find out whether changing his vegetable supplier would increase his takings. He told a friend, Yasmin, who is a statistician: 'It worked. Yesterday using my old supplier my takings were £180, today with the new supplier my takings were £260.' Yasmin persuaded him to carry out a further trial over a two-week period with the following results.

Day	1st week						2nd week					
Day	Mon	Tue	Wed	Thu	Fri	Sat	Mon	Tue	Wed	Thu	Fri	Sat
Supplier	Old	Old	New	Old	New	New	New	New	Old	New	Old	Old
Takings (£)	165	199	215	170	387	408	183	204	221	168	345	389

(a) Using the data from the further trial, apply Wilcoxon's signed-rank test, at the 5% significance level, to investigate whether takings increased when the new supplier was used.

(b) Explain why the conclusion drawn from Jim's original 1-day trial may be invalid and the advantages of the trial designed by Yasmin. Include an explanation of experimental error, replication and randomisation in this context. [A]

10 Eight joints of meat were each cut in half. One half was frozen and wrapped using a standard process and the other half using a new process. The sixteen halves were placed in a freezer and the number of days to spoilage (which can be detected by the colour of the package) was noted for each pack.

Joint number	1	2	3	4	5	6	7	8
Standard process	96	194	149	185	212	237	196	110
New process	117	190	186	776	263	231	242	105
	−	+	−	−	−	+	−	+

A statistician queried the observation on the new process for joint 4. The experimenter agreed that an error must have been made but said that he was certain that, for this joint, the half frozen by the new process had lasted longer than the other half. He had used the sign test on the eight joints and had accepted, at the 5% significance level, that there was no difference in the median number of days to spoilage.

(a) (i) Confirm, by making any necessary calculations, that the sign test applied to these data does lead to the experimenter's conclusion.

   (ii) Use a Wilcoxon's signed-rank test on joints 1, 2, 3, 5, 6, 7 and 8 to test whether there is a difference, at the 5% significance level, in the median number of days to spoilage.

(b) Comment on the validity of using each of the tests on these data. Comment also on the results.

(c) A larger trial is to be carried out and, before the data are collected, you are asked to advise on which test should be used. List advantages of each. [A]

11 An investigation into the effects of alcohol on coordination and processing skills is being carried out.

Twelve people were selected at random from a group of volunteers. They were asked to perform a hand–eye coordination task before and after drinking four units of alcohol.

The times, in seconds, taken to complete the task are given in the following table.

Volunteer	A	B	C	D	E	F	G	H	I	J	K	L
Before alcohol	126	154	129	160	156	153	120	166	134	126	152	163
After alcohol	124	149	142	168	160	159	129	163	141	126	163	166

(a) Carry out a Wilcoxon signed-rank test, at the 5% significance level, to investigate whether the consumption of four units of alcohol has any effect on the median time taken to perform the task. State the null and alternative hypotheses used.

(b) State the assumption, concerning the distribution of times, that is necessary for the Wilcoxon signed-rank test, but not for the sign test, to be valid.

(c) A different suggestion is made for carrying out the investigation.

From the volunteers, 24 people should be selected at random. The first 12 selected perform the task without drinking alcohol and the remaining 12 drink alcohol before performing the task.

Give **one** reason why you think that this suggestion was not used.

12 Students on a statistics course are assessed on coursework and by a written examination. The marks obtained by a sample of 14 students were as follows (3 of the students failed to hand in any coursework):

Student	A	B	C	D	E	F	G	H	I	J	K	L	M	N
% Coursework	68	66	0	65	0	66	69	68	70	67	0	67	69	68
% Examination	53	45	67	52	43	71	37	43	68	27	34	79	57	54
	+	+	−	+	−	−	+	+	+	+	−	−	+	+

(a) Use the sign test, at the 5% significance level on all these results, to examine whether coursework marks are on average higher than examination marks.

(b) Comment on the usefulness or otherwise of these coursework marks as a means of assessing students.

(c) Repeat (a), excluding the three students who failed to hand in coursework.

(d) Summarise, briefly, your conclusions from the previous three parts. [A]

13 A university department is deciding which of two research proposals to support. It asked 11 members of staff to read the proposals and to award each of them a mark out of 100. The marks awarded were as follows:

Member of staff	1	2	3	4	5	6	7	8	9	10	11
Proposal 1	89	37	70	21	29	36	11	46	74	47	26
Proposal 2	95	49	69	86	30	99	19	52	30	45	80

(a) Use the sign test at the 5% significance level to test whether proposal 2 is better than proposal 1.

(b) What assumption was it necessary to make in order to carry out the test in **(a)**?

In view of the erratic nature of the marks awarded, it was decided to ask a further 10 members of staff to award the proposals marks out of 100, but these staff were trained in the features to look for and the method of awarding marks. The new assessment gave the following results:

Member of staff	12	13	14	15	16	17	18	19	20	21
Proposal 1	53	46	60	53	66	59	52	67	63	46
Proposal 2	75	67	69	51	76	59	65	68	72	59

(c) Use a Wilcoxon's signed-rank test, at the 5% significance level, to test whether there is any difference in the average marks awarded to the two proposals by members of staff who had received training.

(d) What assumption is required for **(c)** that is not required for **(a)**?

(e) How would the tests have been affected if staff had not been asked to give the proposals a score but had only been asked to say which of the two proposals they thought was better?

[A]

Key point summary

1 A **distribution-free** or **non-parametric** test does not require the knowledge or assumption that the data involved is normally distributed.   *p164*

2 The **sign test** involves allocating a + or − sign to each reading and using a binomial model with $p = \frac{1}{2}$ to determine the critical region. Usually, the cumulative binomial tables can be used.   *p164*

3 Alternatively, a sign test can be carried out by examining the relevant probability, p, of obtaining the test statistic or a more extreme value. If p is smaller than the stated significance level, then H_0 is rejected. Otherwise, H_0 is accepted. This is a simpler way to carry out the sign test and is acceptable in the exam. *p166*

4 The **Wilcoxon signed-rank test** examines the signed differences between each reading and the suggested population mean or median. Rank order values are then assigned to the differences and, for a two-tailed test, the smaller of the totals T^+ or T^- is the test statistic to be compared with the critical value given in Table 10. If there are 'tied ranks' or equal differences, then the average of the relevant rank values is applied to each reading. *p172*

5 The **Wilcoxon signed-rank test** requires a symmetrical distribution. *p172*

6 The **Wilcoxon signed-rank test** takes the relative magnitudes of the differences into account and is therefore preferred to the **sign test** provided numerical differences can be obtained. *p172*

7 **Experimental design** is used to eliminate **bias** and reduce **experimental error** in data collection *p179*

8 A **paired comparison** is a simple experimental design which can be used to reduce experimental error when two treatments are being compared. In a **paired comparison**, any bias due to differences in the two groups is minimised. *p180*

9 The **sign test** and the **Wilcoxon signed-rank test** are **distribution-free** or **non-parametric tests** which can be used to test for differences between **paired** data. *p184*

In the **sign** and the **Wilcoxon tests**, if a pair of readings are identical, that data pair cannot be used in the evaluation of the test statistic. For tied ranks, the average of the relevant values is assigned to each reading.

8

Test yourself	What to review
1 What is the name of the type of hypothesis tests for which it is not necessary to assume that the data are normally distributed?	*Section 8.1*
2 Give a reason why the Wilcoxon signed-rank test may be preferred to the sign test when testing whether a population median takes a particular value.	*Section 8.2*
3 Which distribution-free test can be used when the data involved is not numerical?	*Section 8.1*
4 For the single sample sign test for a population median, what value of p is used in the binomial model $B(n, p)$?	*Section 8.1*
5 A sign test is carried out on a sample with 14 valid readings. The sign $+$ is allocated to readings above the suggested median for the population and the sign $-$ to those below this median value. Find the critical region for: **(a)** a two-tailed test at 5% significance level, **(b)** a one-tailed test, looking for an increase, at 5% significance level, **(c)** a two-tailed test at the 1% significance level.	*Section 8.1*
6 In a Wilcoxon signed-rank test, the following signed differences are obtained: $\quad$ -4 $\quad$ 6 $\quad$ -2 $\quad$ 1.5 $\quad$ -4 $\quad$ -3 $\quad$ 3 $\quad$ 5 $\quad$ 6 $\quad$ 4 $\quad$ 0.5 $\quad$ 7. Give the signed-rank values which would be allocated to the differences above.	*Section 8.2*
7 In a survey of rush-hour motorists, the distances, in kilometres, travelled to work by a random sample of motorists, who were travelling alone, were: $\quad$ 14 $\quad$ 43 $\quad$ 17 $\quad$ 52 $\quad$ 8 $\quad$ 22 $\quad$ 25 $\quad$ 48 $\quad$ 32 $\quad$ 26 $\quad$ 44. Test the hypothesis, at the 5% significance level, that the median distance travelled to work by motorists travelling alone is 25 km, using: **(a)** the sign test, **(b)** the Wilcoxon signed-rank test. Comment on your results.	*Sections 8.1 and 8.2*
8 When a sign test or Wilcoxon signed-rank test is carried out on a set of paired data and one pair is such that each reading is the same should you: **(a)** include a difference of 0 and allocate a $+$ sign, **(b)** include a difference of 0 and allocate a $-$ sign, **(c)** ignore the pair completely?	*Section 8.4*
9 Under what circumstances can a sign test be carried out on paired data when a Wilcoxon signed-rank test is not possible?	*Section 8.5*

10 To test whether the air temperature at the top of a high tower   *Sections 8.4 and 8.5*
is lower than the temperature at its base, air temperatures
are measured at both top and base at noon on 8 successive
days in June. The results (°C) are given below:

Day	1	2	3	4	5	6	7	8
Top	22	25	12	23	15	19	20	16
Base	24	29	16	21	18	23	21	20

(a) Use a sign test at the 5% level of significance to test
whether the median temperature at the top of the tower
is lower than that of the temperature at its base.

(b) Use a Wilcoxon signed-rank test to test the same hypothesis
as stated in (a) at the same level of significance.

(c) Comment on your results from (a) and (b) and give a
reason why the Wilcoxon signed-rank test might be
preferred to the sign test.

---

1 Distribution-free or non-parametric.

2 Takes into account rank order of differences – not just the sign of the differences.

3 Sign test.

4 $p = 0.5$.

5 (a) $\{0,1,2,12,13,14\}$ probability < 0.05 [$\{0,1,2,3,11,12,13,14\}$ has probability closest to 5%, 0.05 (but bigger)];

(b) $\{11,12,13,14\}$;

(c) $\{0,1,13,14\}$ probability < 0.01 [$\{0,1,2,12,13,14\}$ has probability closest to 1%, 0.01 (but bigger)].

6 $^{-}7, ^{+}10\frac{1}{2}, ^{-}3, ^{+}2, ^{-}7, ^{-}4\frac{1}{2}, ^{+}4\frac{1}{2}, ^{+}9, ^{+}10\frac{1}{2}, ^{+}7, ^{+}1, ^{+}12$.

7 (a) $\mathbf{H_0}$ pop med $= 25$, $\mathbf{H_1}$ pop med $\neq 25$, ts $= 6^{+}$ or 4^{-} $n = 10$,   5% two-tail.
$P(\geqslant 6^{+}) = P(\leqslant 4^{-}) = 0.377 > 2.5\%$.
Accept $\mathbf{H_0}$. No significant evidence to doubt pop med $= 25$.

(b) $\mathbf{H_0}$ pop med $= 25$, $\mathbf{H_1}$ pop med $\neq 25$, $T^{+} = 38$, $T^{-} = 17$,   5% two-tail.
$n = 10$,   cv $= 8$,   ts $= 17$.
Accept $\mathbf{H_0}$. No significant evidence to doubt pop med $= 25$.
In this case both tests lead to the same conclusion.

8 (c).

9 When the sign of each difference is known but it is not possible to rank their magnitudes or if the population is known not to be symmetrical.

10 (a) $P(1 \text{ or fewer}+) = 0.0352$, significant evidence that median temperature is lower at the top;

(b) $T = 2.5$, cv $= 6$, significant evidence that median temperature is lower at the top;

(c) In this case both tests give the same result. In general, because Wilcoxon's test uses the magnitudes as well as the signs of the differences, it is more likely to detect a difference if one exists.

CHAPTER 9

Distribution-free methods for two or more unpaired samples

Learning objectives

After studying this chapter, you should be able to:

- understand the difference between paired and unpaired data
- carry out a Mann–Whitney U test on data collected as two unpaired samples
- carry out a Kruskal–Wallis test.

9.1 The Mann–Whitney U test

When it is not possible to gather paired data and when you still cannot assume or do not know that the data is normally distributed, a Mann–Whitney U test can be used to test for differences, on average, between the populations from which the two samples are taken. As with other tests, the hypothesis has to be formulated more precisely. In this case it is that the two populations are identical.

> The **Mann–Whitney U test** is a **distribution-free** or **non-parametric test** which can be used to test for differences between two sets of data which are not **paired**.

For example, a hospital consultant may wish to collect information from the patients in his clinic in order to compare a blood measurement for male and female patients. The data is obviously unpaired and the sample sizes are likely to be different.

Or, perhaps, if Julia, who loves to eat grapefruit every day, believes that the grapefruit in one supermarket are larger than those from another supermarket, she would need to obtain two separate samples: one of grapefruit from Supermarket A and another from Supermarket B. Clearly this is not a paired sample, but two unrelated samples of fruit.

These examples of unpaired samples will be used to explain the procedure for the Mann–Whitney U test. This test, like the Wilcoxon, considers rank values. The ranks are of the actual data, not the differences, as the data is not in pairs.

Sometimes this test is called the Wilcoxon rank-sum test.

Worked example 9.1

A cardiac consultant at a large hospital decides to measure the cholesterol levels of a sample of outpatients seen at her clinic. The resultant readings were:

Females	5.3	2.1	3.8	10.0	1.7	7.1	9.8	1.9	3.1	1.5	
Males	7.3	1.8	4.9	9.2	7.1	4.2	10.2	10.9	3.7	10.5	6.2

In this case, the two samples may well be of different sizes. One is called size m, the other size n.

Test, at the 5% significance level, the hypothesis that there is no difference between male and female cholesterol levels.

Solution

There are two separate samples here: males and females and so an unpaired test is required with:

H_0 Average population cholesterol level for females
  = Average population cholesterol level for males
H_1 Average population cholesterol level for females
  ≠ Average population cholesterol level for males

Or
H_0 two samples come from identical populations
H_1 two samples do not come from identical populations: cholesterol levels differ on average.

Two-tailed test significance level 5%.

Ranking the data gives:

The ranking is done with 1 as the lowest.

Females	11	5	8	18	2	13.5	17	4	6	1	
Males	15	3	10	16	13.5	9	19	21	7	20	12

For the females, $T_f = 85.5$ and $m = 10$

For the males, $T_m = 145.5$ and $n = 11$

There are two identical readings of 7.1. One is a female, the other a male. The rank order values for these two readings should be 13 and 14. In order to reflect this situation, the **average** of these two rank values is assigned to each of the readings of 7.1. Therefore, the rank 13.5 appears in each list.

9

The **test statistic** is labelled **U** where

$$U = T - \frac{n(n+1)}{2}$$

T is the sum of the ranks of the sample of size n.

If you have read the note opposite then it is clear that there are *two* possible values for U depending on how the two samples are labelled.

Since it really does not matter which sample is chosen to have size n and which to have size m, you could equally well say:

$$U = T - \frac{m(m+1)}{2}$$

T is the sum of the ranks of the sample of size m.

The two possible test statistics are:

$$U = 85.5 - \frac{10(10 + 1)}{2} = \mathbf{30.5} \text{ using female data}$$

or

$$U = 145.5 - \frac{11(11 + 1)}{2} = \mathbf{79.5} \text{ using male data}$$

> A check can be made as the two possible test statistics should add up to **mn**. In this case:
> $$30.5 + 79.5 = 110$$
> $$mn = (10 \times 11) = 110$$

From the tables, the **lower** critical value for a two-tail test, at a 5% significance level, with samples of size 10 and 11, is **27**.

As the **lower** tail is given, the smaller of the two possible **test statistics** above is the relevant one to use for two-tailed tests.

Thus, with **U** = 30.5, comparing this test statistic to the lower critical value of 27, **U** is **not** less than 27 and therefore **U** does **not** lie in the critical region.

> Note that, alternatively, the upper tail critical value can be evaluated using the formula:
> **Upper tail** = *mn* − **lower tail critical**
> $= (10 \times 11) - 27 = \mathbf{83}$
> Then 79.5 is the relevant test statistics and this is **less** than the upper tail and so there is no significant evidence to doubt **H₀**.
> Both 30.5 and 79.5 lie in between the lower and upper tail critical values of 27 and 83 and so, clearly, **H₀** is accepted as true.

Conclusion

The test statistic **U** = 30.5 is **greater** than the lower critical value of 27. There is no significant evidence to doubt **H₀** and we conclude that there is no evidence of a difference in average cholesterol levels between males and females in the population.

> Note that, theoretically, **H₁** only states that the populations are **not identical**. However, because of the nature of the test, **H₀** is only likely to be rejected if the populations differ on average (whichever measure of average is used).
>
> If the populations had the same mean but very different standard deviations, then **H₀** is **not** likely to be rejected.

Worked example 9.2

Julia has obtained two samples of grapefruit: one from supermarket A and one from supermarket B. These grapefruit were weighed on the same scales and the results obtained for their weights in grams were:

A	164	212	132	140	116	104	167	
B	208	246	197	153	118	169	120	144

Use the Mann–Whitney U test to investigate whether there is any significant evidence, at the 5% level, that the grapefruit from supermarket A are smaller than those from supermarket B.

> The word *smaller* indicates a one-tailed test.

Solution

H_0 Population average weight A = Population average weight B
H_1 Population average weight A $<$ Population average weight B

One-tailed test significance level 5%.

The first step in carrying out a Mann–Whitney U test is to rank all the data in both samples as though they were all in one group. This is illustrated below.

A	9	14	5	6	2	1	10	
B	13	15	12	8	3	11	4	7

The total for the ranks in each group is then found.

> Rank total A, $T_A = 47$, Rank total B, $T_B = 73$

The two possible values for **U** are:

> using sample B, $n = 8$, $T_B = 73$, $\quad U = 73 - \dfrac{8(8 + 1)}{2} = \mathbf{37}$

or

> using sample A, $m = 7$, $T_A = 47$, $\quad U = 47 - \dfrac{7(7 + 1)}{2} = \mathbf{19}$

The critical values are given in Table 11 for one- or two-tailed tests at 5%.

In this case, for a one-tailed test where we are expecting to find the ranking for A overall **lower** than the ranking for B, the lower value for A of **U** = 19 is the relevant test statistic.

The critical value given for $m = 7$, $n = 8$, is 13 for a one-tailed test at 5% and the **test statistic** of 19 is clearly **not lower** than 13.

> **Or**
> H_0 two samples come from identical populations.
> H_1 two samples do not come from identical populations: population A has a lower average weight.

> Again, check that the two possible test statistics add up to *mn*.
> In this case,
> $37 + 19 = 7 \times 8 = 56$.

> If you take a minute to look at Table 11, you will see that the critical values given are symmetrical.
> For $m = 7$ and $n = 8$ the critical value for a one-tailed test is 11 but the critical value is also 11 when $m = 8$ and $n = 7$.

> The **critical region** includes the value quoted in Table 11. A **test statistic** $\leqslant 13$ leads to rejection of H_0.
> The critical values given in the tables are those with significance levels closest to those stated.

Conclusion

The test statistic of 19 is **not less** than the quoted **lower tail critical value** so there is no significant evidence to doubt H_0. We conclude that there is no significant evidence to suggest the population weights differ on average.

9.2 The Mann–Whitney U test procedure

1 Put **all** data in rank order as though it was in **one** group only.

2 Assign rank 1 to the **smallest** value.

3 Find the **total, T**, of the rank values in each sample.

4 Find the test statistic **U**, where

$$U = T - \frac{n(n+1)}{2}$$

There are **two** possible values for **U**, one from each sample.

5 Compare appropriate **U** with relevant critical values from Table 11.

> All the data, in both samples, is **put in rank order** as though it was in one big group. The smallest value is given rank value 1. The total of the rank values in each sample, say, T_A and T_B, are found.

> Two possible values exist for the test statistic, **U**, where:
>
> $$U = T - \frac{n(n+1)}{2}$$
>
> for T, the total of the rank values in the sample of size n.

Worked example 9.3

A French teacher has decided to try a new approach in getting his students to learn vocabulary. He selects two groups each of six students of similar ability. One group is left to learn their vocabulary in the usual way but the other group is instructed in the new approach.

The two groups are then given a test and the number of words correctly remembered out of the list of 30 is given below:

Usual way	14	17	17	11	16	15
New approach	24	25	29	18	16	19

(a) In the context of this experiment, which group is the control group?

(b) Is the data supplied paired data or unpaired data?

(c) Comment on the experimental design used by this teacher.

(d) Test the hypothesis that the new approach leads to a higher success rate in learning French vocabulary. Use a 5% significance level.

> The use of the word **higher**, indicates a one-tail test.

Solution

(a) The group which is left to learn the vocabulary in the usual way is the control group.

(b) The data is made up of two quite separate groups so it is unpaired data.

> Although the two samples are of the same size, the data is **not** paired.

(c) It would not be possible to test the same person twice, on the same vocabulary list, using both methods, so a paired test is not possible and, as the teacher has ensured the two groups are well matched, hopefully for gender, ability and motivation, the design of this experiment is good.

(d) For number of words in vocabulary list:

H_0 Population average correct = Population average correct
　　　for usual method 　　　　　for new approach
H_1 Population average correct < Population average correct
　　　for usual method 　　　　　for new approach

One-tailed test significance level 5%.

Ranks are as follows:

Usual way	2	6.5	6.5	1	4.5	3
New approach	10	11	12	8	4.5	9

For the usual way, 　　　$T_u = 23.5$ and $m = 6$
For the new approach, 　$T_n = 54.5$ and $n = 6$.

> There are two occasions with this data when readings are repeated. There are two readings of 16, one in the results from the usual and one from the new method. They share the ranks 4 and 5 and are given an average rank of 4.5 each. Also, there are two readings of 17 in the usual results. These share the rank values 6 and 7 and it doesn't matter which is allocated 6, which 7 or whether they each have rank 6.5 (as given). The total T_u will still be 23.5.

9

The two possible test statistics are:

$$U = 23.5 - \frac{6(6+1)}{2} = 2.5 \text{ using usual results}$$

or

$$U = 54.5 - \frac{6(6+1)}{2} = 33.5 \text{ using new method results}$$

> As you become used to this test, you will not need to find both possible test statistics. For a one-tailed test only one value of U is relevant and for a two-tailed test it will usually be obvious which value will be the smaller.

In this case, for a one-tailed test where we are expecting to find the ranking for the usual way results overall **lower** than the ranking for the results using the new method, the value of for the usual method, $U = 2.5$, is the relevant test statistic to compare with the lower tail critical value.

The critical value given for $m = 6$, $n = 6$, is 7 for a one-tailed test at 5% and the **test statistic** of 2.5 is clearly **lower** than 7.

Conclusion

Since the test statistic 2.5 is much **lower** than the lower tail critical value of 7, there is significant evidence that the two populations differ and students achieve a higher number of correct answers for vocabulary tests using the new approach.

9.3 Comparison of U with critical values

Table 11 gives **lower tail** critical values, therefore, as seen in the previous examples, it is easier, to consider the **lower U** value only.

In **one-tailed tests**, consider which of the two **U** test values is expected to be small and this **lower U** is compared with the **lower, one-tail, critical** value given in Table 11.

In **two-tailed tests**, the **lower U** is compared with the **lower, two-tail, critical** value given in Table 11.

H_0 is rejected if **lower U** $\leqslant$ one-tail **critical value**.

H_0 is rejected if **lower U** $\leqslant$ two-tail **critical value**.

 Table 11 tabulates the **lower** tail critical values.

 In **two-tailed** tests, it is only necessary to compare the lower value of U with the **lower tail** critical value from Table 11. If this **lower** test value is greater than the critical value, then both possible U test values will lie between the upper and lower critical values.

 In one-tailed tests H_1, the alternative hypothesis, will determine which value of U should be compared with the lower tail critical value. This will usually, but not always, be the lower value of U.

EXERCISE 9A

1 Twenty students were selected at random from a year group of 121 in a large school. The students were all asked to carry out a visual test and their reaction times(seconds) were noted:

Girls	6.6	7.2	13.1	7.6	4.3	6.7	9.5	3.6	1.4	
Boys	19.7	11.8	7.5	3.0	23.3	6.4	14.1	6.0	15.4	3.8 6.9

Carry out a Mann–Whitney U test at the 5% significance level to test whether girls have faster reaction times than boys for this test.
Comment critically on the design of this experiment.

2 Car batteries are sold as either 'standard' or, for a higher price, as premium 'longer-lasting'. A car hire company decides to test whether the premium batteries do indeed have a longer life. Seven of their small saloon cars are fitted with the standard battery and seven have the premium one fitted.

The length of life (months) of all 14 batteries is given below:

Standard	21	28	37	23	43	40	45
Premium	53	40	48	60	36	51	46

Carry out a Mann–Whitney U test at the 5% significance level to determine whether the premium batteries do last longer. What bias might exist in the way the data was collected for this experiment?

3 A town fair ran an archery competition for children aged between 12 and 15 years. There were 12 entrants for the competition. Four were girls and eight boys. Their overall scores at the end of the competition were:

Girls	865	975	685	785				
Boys	990	780	955	785	970	985	750	835

At the 5% level of significance, use a suitable non-parametric test to determine whether there is any significant difference between the scores for boys and girls at archery.

9

4 Applicants for a precision job at a component manufacturer are screened at interview by taking a test to judge their suitability. Two different tests are used randomly for these applicants but the staff recruitment manager is concerned that one test is more difficult than the other.

Results from the 15 most recent tests (marked out of 40) were as follows:

Test 1	23	32	28	29	33	27	30	
Test 2	33	35	37	27	31	28	36	32

Use a Mann–Whitney U test to determine whether there is any evidence, at the 5% significance level, that applicants score lower marks on Test 1.

Explain, with reference to this question, what is meant by *experimental error*.

5 A botanist takes samples of a particular heather from both sides of a hillside. One side faces east and the other side faces west. He measures the length of a random selection of new season plants to examine the hypothesis that plants on the west facing side grow faster than those on the east facing side. The lengths (mm) were found to be as follows:

West	42	29	34	37	35	28	32	28	36
East	38	26	30	27	25	28	31		

Carry out a suitable non-parametric test to test the botanist's hypothesis at the 5% significance level.
Explain your choice of test with reference to the data collected by the botanist. [A]

6 The vitamin content of the flesh of each of a random sample of eight oranges and of a random sample of five lemons was measured. The results are given in milligrams per 10 grams.

Oranges	1.14	1.59	1.57	1.33	1.08	1.27	1.43	1.36
Lemons	1.04	0.95	0.63	1.62	1.11			

Carry out a suitable non-parametric test to investigate at the 5% significance level, whether the average vitamin content of lemons is lower than that of oranges. [A]

7 An economist believes that a typical basket of weekly provisions, purchased by a family of four, costs more in Southville than it does in Nortown. Six stores were randomly selected in each of these two cities and the following costs, for identical baskets of provisions, were observed.

Southville	12.32	13.10	12.11	12.84	12.52	12.71
Nortown	11.95	11.84	12.22	12.67	11.53	12.03

(a) Explain why a paired test would not be appropriate here.

(b) Carry out a Mann–Whitney U test at the 5% significance level to determine whether there is any difference, on average, between the cost of the basket in Southville and Nortown. [A]

8 The manager of a road haulage firm records the time taken on six occasions for a lorry to travel from the depot to a particular customer's factory. Roadworks are due to start on the usual route so the manager decides to try an alternative route and records the times of eight journeys on this new route.

Time (minutes), old route	34	45	36	48	49	38		
Time (minutes), alternative route	43	35	47	39	58	40	39	51

(a) Use a suitable distribution-free test to investigate, at the 5% significance level, whether there is a difference in the average time taken on the two routes.

(b) One driver had taken 99 minutes on the alternative route. Investigation showed that this was due to losing his way and it was decided to exclude this result from the tests. Comment on this decision. [A]

9 A firm is to buy a fleet of cars for use by its representatives and wishes to choose between two alternative models, A and B. It places an advertisement in a local paper offering four free gallons of petrol to anyone who has bought a new car of either model in the last year. The offer is conditional on being willing to answer a questionnaire and to note how far the car goes, under typical driving conditions, on the free petrol supplied. The following data were obtained:

Miles driven on four gallons of petrol

Model A	117	136	108	147					
Model B	98	124	96	117	115	126	109	91	108

(a) Use the Mann–Whitney U test to investigate, at the 5% significance level, whether there is a difference between the average petrol consumption of the two models.

(b) List good and bad features of the experimental method. [A]

10 A wholesale fruiterer stocks two varieties of oranges, P and Q, for distribution to retail outlets.

Inspector A measured the weight, x grams, of each of a random sample of 10 oranges of variety P with the following results.

155 146 149 156 161 158 144 148 165 158

A second inspector, B, measured the weight, y grams, of each of a random sample of 11 oranges of variety Q with the following results.

149 146 151 142 153 144 148 152 140 152 137

(a) Indicate a possible source of bias in the collection of these data and suggest how it could have been avoided.

(b) Use a suitable distribution-free test to investigate, at the 5% significance level, the hypothesis that there is no difference between the average weights of the two varieties of oranges. [A]

11 The manager of a bicycle shop wishes to compare the performance of two makes of tyre. She advertises that anyone buying a tyre of either make during the next week will be given a free inner tube. However they must agree to note the length of time from the new tyre being put on until it sustains a puncture.
The following data were collected.

Make of tyre	Number of days to puncture					
A	43	155	3	167	212	142
	12	96	78	191	77	
B	23	144	34	22		

(a) Use a suitable non-parametric test to investigate, at the 5% significance level, whether the average number of days before sustaining a puncture is the same for the two makes of tyre.

(b) Comment on the method of data collection, the appropriateness of the variable measured, the sample sizes and on any other matters relevant to the experiment.

(c) Explain in the context of this experiment the meaning of Type I and of Type II errors. [A]

12 The development engineer of a company making razors records the time it takes him to shave, on six mornings, using a standard razor made by the company. The times, in seconds, were

217   208   254   237   232   243.

(a) Assuming that this may be regarded as a random sample of his shaving times test, at the 1% significance level, the hypothesis that the median time he takes to shave is more than 3 and a half minutes. Use either the sign test or Wilcoxon's signed-rank test, whichever is more appropriate. Justify your choice of test.

He wishes to compare the time taken by different designs of razor. He decided that rather than test all designs himself it would be quicker to find other employees who would be willing to test one design each. As a preliminary step his assistant agrees to test the standard razor and produces the following times:

$$186 \quad 219 \quad 168 \quad 202 \quad 191 \quad 184.$$

(b) Use a Mann–Whitney U test to investigate at the 5% significance level whether the average shaving times of engineer and assistant are the same.

(c) Advise the engineer on how to proceed with his investigation. [A]

13 Two analysers are used in a hospital laboratory to measure blood creatinine levels. These are used as a measure of kidney function.

To compare the accuracy of the measurements taken by the two machines a technician took eight specimens of blood and measured their creatinine level (in micromoles per litre) using analyser A. She then measured a further seven samples using analyser B.
The results were as follows:

Analyser A	103	124	130	151	108	96	103	121
Analyser B	142	139	156	164	142	119	117	

(a) Use the Mann–Whitney U test, and a 5% significance level, to investigate for differences between the results of the two analysers.

The technician then realised that, as the analysers had been measuring different samples, whatever the result of the test in **(a)**, no conclusion about their accuracy could be drawn. She then took eight specimens of blood and measured the creatinine level of each specimen using each machine. The results were as follows:

Specimen 1	1	2	3	4	5	6	7	8
Analyser A	119	173	100	99	77	121	84	73
Analyser B	106	153	83	95	69	123	84	67

(b) Carry out Wilcoxon's signed-rank test, at the 5% level, to compare the performance of the two machines.

(c) Is it now possible to decide which machine is more accurate?
Explain your answer.

9

A statistician requested that each analyser should be used repeatedly to analyse a standard solution which should give a creatinine level of 90. The results for each machine were as follows:

Analyser A	93	94	96	91	89	93	95	93
Analyser B	98	107	83	85	94	109	89	92

(d) Use Wilcoxon's signed-rank test to test, for each analyser, the hypothesis that the mean measurement is 90. Use a 5% significance level.

(e) What can now be concluded about the performance of the two machines,

 (i) from the conclusions reached in (d),

 (ii) from examining (without undertaking further hypothesis testing) the variability of the most recent data? [A]

14 Random samples of apples of two different varieties held in a warehouse were weighed. The weights in grams were as follows:

Variety 1	110.5	89.6	115.0	98.2	113.1	104.3	85.6	92.0		
Variety 2	125.6	118.3	118.0	110.8	116.5	108.7	108.2	104.4	114.4	98.4

(a) Use the Mann–Whitney U test to investigate, at the 5% significance level, whether apples of variety 2 are on average heavier than the apples of variety 1.

(b) Later it transpired that the measuring device used to weigh the apples was faulty and that all the apples were 10 grams heavier than the recorded weight. How would this further information affect your conclusions?

(c) If the recorded weight for variety 1 was correct but variety 2 apples were 10 grams heavier than recorded, state, without further calculation, how your conclusions would be affected. [A]

9.4 The Kruskal–Wallis test

> The **Kruskal–Wallis** test is the distribution-free test which is used to test for differences between three or more populations.

When independent samples from three or more populations are to be compared, the Kruskal–Wallis test can be used to test for differences, on average, between the populations from which the samples are taken. It does not require the populations to be known or assumed to be normally distributed. As with other tests, the null hypothesis has to be formulated precisely. In this case, it is that the populations are identical. However, because of the nature of the test, only a difference in 'average' is likely to lead to a rejection of H_0 and so the test is regarded as a test of average. This is exactly the same as the Mann–Whitney test.

Hypotheses which refer to the population and a measure of average will be marked correct.

The alternative hypothesis is that at least two of the populations must differ. Further testing, following a rejection of the null hypothesis, is required to establish the exact nature of the difference or differences identified.

This further testing will not be expected in the exam.

For example, a psychologist may wish to collect information from children in order to compare the tendency to traditional sex roles for children whose mothers work full-time, part-time or who do not work at all outside the home. The data is obviously unpaired, there are three groups and the sample sizes are likely to be different.

This example of more than two unpaired samples will be used to explain the procedure for the Kruskal–Wallis test. This test, like the Mann–Whitney, does not require the groups to be of equal size and it considers rank values of the actual data, ranked together as though they were one group.

Worked example 9.4

A psychologist collects information from children in order to compare the tendency to traditional sex roles for children whose mothers work full-time, part-time or who do not work at all outside the home. The children were given several scenarios and were asked to comment. The children were given scores out of a maximum of 100, where 100 indicated the most extreme sex role stereotyping.

9

The scores for the three independent groups of children involved were as follows:

Full time	16	27	32	34	26	31	
Part time	22	45	60	37	48	54	
Do not work	20	64	79	30	40	61	78	82

Note the three medians:

$\eta_F = 29$

$\eta_P = 46.5$

$\eta_D = 62.5$

Use the Kruskal–Wallis test to investigate whether there is any significant evidence, at the 5% level, that the median sex role stereotyping scores differ for the three groups.

Solution

H_0 samples taken from identical populations
H_1 at least two of the population averages differ or samples taken from populations which are not identical
5% significance level.

Or

H_0 $\eta_F = \eta_P = \eta_D$

H_1 at least two of η_F, η_P, η_D differ.

Note that the acceptance of the alternative hypothesis only concludes that **at least two** of the populations differ. All three may differ, or just two.

This is illustrated below.

Full time	1	5	8	9	4	7		
Part time	3	12	15	10	13	14		
Do not work	2	17	19	6	11	16	18	20

The total for the ranks in each group is then found.

Rank total: full time $\quad T_F = 34$
Rank total: part time $\quad T_P = 67$
Rank total: do not work $\quad T_D = 109$

$N = 20$ so $\frac{1}{2}N(N + 1) = 210$

$$\sum_{i=1}^{k} \frac{T_i^2}{n_i} = \frac{34^2}{6} + \frac{67^2}{6} + \frac{109^2}{8} = 2425.958$$

Hence, $H = \dfrac{12}{20 \times 21} \times 2425.958 - 3 \times 21$

$\qquad = \dfrac{1}{35} \times 2425.958 - 63$

$\qquad = 6.31$ (to 3 s.f.)

If the null hypothesis is true and the samples are reasonably large the distribution of H may be approximated by the χ^2 distribution with $k - 1$ degrees of freedom (where k is the number of samples). It is generally accepted that if all samples contain at least five observations the χ^2 distribution is a good approximation. Tables of critical values exist for use when the samples are smaller but these will not be required in this unit.

 The critical values are obtained from Table 6: Percentage points of the χ^2-distribution. The degrees of freedom are $k - 1$, where k is the number of populations to be compared.

In this case, for a test at 5% with three populations to compare, so two degrees of freedom, the critical value is 5.99. The Kruskal–Wallis test always requires the one-tailed χ^2 critical value.

The **test statistic** of 6.31 is higher than 5.99.

You will not be asked to rank more than three samples in the exam. If a question involves more than three samples, some or all of the rank values will be given.

The first step in carrying out a Kruskal–Wallis test is to rank all the data in all samples as though they were all in one group.

The total number of readings given, $N = 20$.

A check can be made, as with the Mann–Whitney rank totals.

$\sum T_i = \frac{1}{2}N(N + 1)$

Here, $34 + 67 + 109 = 210$.

Test statistic

$$H = \frac{12}{N(N + 1)} \sum_{i=1}^{k} \frac{T_i^2}{n_i} - 3(N + 1)$$

The formula for this test statistic, H is given in the AQA booklet of formulae and tables.

The value of H is the same whether rank 1 is given to the highest score or to the lowest score.

This is similar to contingency tables where the distribution of X^2 is approximated by χ^2.

A test statistic $\geqslant 5.99$ leads to rejection of $\mathbf{H_0}$.

Conclusion

There is significant evidence to doubt H_0. We conclude that there is significant evidence to suggest that at least two of the populations differ, or at least two of the population medians differ.

It would certainly appear that there is a significant difference in the population median scores for children whose mothers work full-time and those whose mothers do not work at all.

> The rejection of H_0 leads to the conclusion that at least two of the populations differ.

Note that the rejection of H_0 only concludes that **at least two** of the populations differ. All may differ, or just two.

Worked example 9.5

Three different types of milling machine are being considered for purchase by a small engineering company. In order to assess these machines, each type was obtained on free loan and was randomly assigned to one of 16 technicians, all equally skilled in machine operation.

Each machine was put through a series of tasks and was assessed in various categories. A total score, out of a maximum of 40, was then assigned to each machine by the technicians.

The scores are given in the following table:

Machine 1	Machine 2	Machine 3
24.4	28.3	32.1
23.3	34.1	34.2
26.3	32.2	36.1
27.0	30.0	32.4
29.8	29.4	35.5
		33.9

Note the three medians:

$\eta_1 = 26.3$

$\eta_2 = 30.0$

$\eta_3 = 34.05$

9

(a) Perform a Kruskal–Wallis test to investigate whether there is any significant difference in median scores between the machines. Use the 1% significance level.

(b) Recommend to the company which type of machine should be purchased.

Solution

(a) H_0 samples taken from identical populations
H_1 at least two of the population averages differ or samples taken from populations which are not identical
1% significance level

Or

H_0 $\eta = \eta_2 = \eta_3$

H_1 at least two of the η_1, η_2, η_3 do differ.

Ranks as one group:

Machine 1	Machine 2	Machine 3
15	12	8
16	4	3
14	7	1
13	9	6
10	11	2
		5

The total number of readings given, $N = 16$.

The total for the ranks in each group is then found.
Rank total: Machine 1, $T_1 = 68$
Rank total: Machine 2, $T_2 = 43$
Rank total: Machine 3, $T_3 = 25$

A check can be made, as with the Mann–Whitney rank totals.
$\sum T_i = \frac{1}{2}N(N + 1)$
Here, $68 + 43 + 25 = 136$.
$N = 16$, so $\frac{1}{2}N(N + 1) = 136$.

Test statistic $H = \dfrac{12}{N(N + 1)} \displaystyle\sum_{i=1}^{k} \dfrac{T_i^2}{n_i} - 3(N + 1)$

$$\sum_{i=1}^{k} \frac{T_i^2}{n_i} = \frac{68^2}{5} + \frac{43^2}{5} + \frac{25^2}{6} = 1398.77$$

Hence, $H = \dfrac{12}{16 \times 17} \times 1398.77 - 3 \times 17$

$\qquad = \dfrac{3}{68} \times 1398.77 - 51$

$\qquad = 10.7$ (to 3 s.f.)

In this case, for a test at 1% with three populations to compare, so two degrees of freedom, the critical value is 9.21.

$\qquad 10.71 > 9.21$

Conclusion

There is significant evidence to doubt $\mathbf{H_0}$. We conclude that there is significant evidence to suggest that at least two of the populations differ, or at least two of the population medians differ.

(b) It would certainly appear that there is a significant difference in population median scores for Machine 1 and Machine 3. Since Machine 3 is preferable to Machine 1 and Machine 2 has a median score between these two, then Machine 3 is recommended for purchase.

Worked example 9.6

A chemist is working on the use of synthetic pheromones (hormones involved in mating) as a bait in a trap to catch unwanted insects. Four different levels of the synthetic hormone are used in traps placed at random in a peach orchard. The numbers of insects trapped in each trap during a four hour period are given in the following table.
Some of the rank values (where smallest number has rank value 1) for these numbers are also given.

Level 1 number	rank	Level 2 number	rank	Level 3 number rank	Level 4 number rank
2	1	22		24	18
4	3	12	8	25	54
10	6	11	7	36	26
3	2	17		16	76
5	4	6	5	38	33

Note the four medians:
$\eta_1 = 4$
$\eta_2 = 12$
$\eta_3 = 25$
$\eta_4 = 33$

(a) Perform a Kruskal–Wallis test to investigate whether there is any significant difference in median numbers of trapped insects between the four different levels of pheromone used. Use the 1% significance level.

(b) Interpret your conclusion in context.

Solution

(a) H_0 samples taken from identical populations
H_1 at least two of the population averages differ or samples taken from populations which are not identical

5% significance level

Ranks as one group.

Or
H_0 $\eta_1 = \eta_2 = \eta_3 = \eta_4$
H_1 at least two of $\eta_1, \eta_2, \eta_3, \eta_4$ do differ.

Level 1	Level 2	Level 3	Level 4
1	12	13	11
3	8	14	19
6	7	17	15
2	10	9	20
4	5	18	16

The total number of readings given, $N = 20$.

The total for the ranks in each group is then found.
Rank total: Level 1 $T_1 = 16$
Rank total: Level 2 $T_2 = 42$
Rank total: Level 3 $T_3 = 71$
Rank total: Level 4 $T_4 = 81$

A check can be made, as with the Mann–Whitney rank totals.
$\sum T_i = \frac{1}{2}N(N + 1)$
Here, $16 + 42 + 71 + 81 = 210$.
$N = 20$, so $\frac{1}{2}N(N + 1) = 210$.

Test statistic $H = \dfrac{12}{N(N + 1)} \sum_{i=1}^{k} \dfrac{T_i^2}{n_i} - 3(N + 1)$

$\sum_{i=1}^{k} \dfrac{T_i^2}{n_i} = \dfrac{16^2}{5} + \dfrac{42^2}{5} + \dfrac{71^2}{5} + \dfrac{81^2}{5} = 2724.4$

Hence, $H = \dfrac{12}{20 \times 21} \times 2724.4 - 3 \times 21$

$= 14.8$ (to 3 s.f.)

In this case, for a test at 1% with four populations to compare, so three degrees of freedom, the critical value is 11.345.

$14.8 > 11.345$

9

Conclusion

There is significant evidence to doubt H_0. We conclude that there is significant evidence to suggest that at least two of the populations differ, or at least two of the population medians differ.

(b) It would certainly appear that there is a significant difference in population median numbers for Level 1 and Level 4. Level 4 of the pheromone appears the most effective at trapping insects.

EXERCISE 9B

1 Sixteen children were selected at random from the same year group in a large school. The children were randomly allocated to be taught under one of three conditions: high, medium or low motivation. The reading scores for these children are given in the following table.

High motivation	Medium motivation	Low motivation
18	9	3
20	11	2
23	8	5
21	12	1
19	10	7
		6

(a) Carry out a Kruskal–Wallis test at the 1% significance level to test whether the median reading scores differ for the three motivation levels.

(b) Interpret your conclusion in context.

2 Batteries are sold as either 'Bargain', 'Premium' or 'Long-life'. A company decides to test whether there is any difference in average life length for the three types of batteries.
Twelve identical calculators are randomly selected from those used each day by staff in two departments of the company. The calculators are then randomly allocated to either a 'Bargain', a 'Premium' or a 'Long-life' battery.
The length of life (hours) of all twelve batteries is given in the following table.

Bargain	Premium	Long-life
112	142	152
135	122	145
101	106	109
96	103	105

(a) Carry out a Kruskal–Wallis test at the 5% significance level to determine whether there is any difference in average length of life for the three types of batteries.

(b) What bias might exist in the way the experiment was carried out?

3 A psychologist wishes to investigate the level of absenteeism for a group of patients suffering from depression. Each patient is randomly allocated to receive either a placebo (a tablet containing no active ingredient), an existing drug or a new drug.

The number of days of absence from work recorded for each patient in the trial over a four week period are given in the following table.

Placebo	Existing drug	New drug
12	7	3
13	8	5
9	6	11
10	14	4

(a) At the 5% level of significance, use a suitable distribution free (non-parametric) test to determine whether there is any significant difference between the average absenteeism level for the existing drug, the new drug or the placebo.

(b) Interpret your conclusion in context.

4 An animal behaviourist is investigating evidence for learning in mice. Sixteen mice are divided into four groups. Each group is given a different amount of experience in a maze.

The mice in group A have one experience, those in group B have two experiences, those in group C have three experiences and those in group D have four experiences.

Each mouse is then placed in a maze and the time, in minutes, taken to solve the maze is recorded. The results are given in the following table, together with some of the rank values.

Group A		Group B		Group C		Group D	
time	rank	time	rank	time	rank	time	rank
19		14		10	5	6	1
20		15		9	4	8	3
17		16		13		7	2
21		19		12		11	6

Carry out a Kruskal–Wallis test at the 5% significance level to investigate whether there is any significant difference in the average times taken to solve the maze by the four groups of mice.

Interpret your conclusion in context.

5 Three special ovens in a metal working shop are used to heat metal specimens. All the ovens are supposed to operate at the same temperature. It is known that the temperature of an oven varies, and it is suspected that there are significant mean temperature differences between ovens. The table below shows the temperatures, in degrees centigrade, of each of the three ovens on a random sample of heatings.

Oven	Temperature °C					
1	494	497	481	496	501	499
2	488	490	479	478	475	
3	489	483	467	472	474	

Use the Kruskal–Wallis test, at the 10% significance level, to examine whether there is a difference in average temperature for the three ovens. Interpret your results in context.

6 Fifty-two people took part in a fun run to raise money for an international charity. On finishing the run each participant was asked which sporting activity, apart from running, they participated in most. The table below shows their order of finishing the fun run classified by their reply to the question.

Cricket	12	18	29	31	46					
Darts	9	37	40	44	48	50				
Football	2	5	6	23	24	28	32	38	42	43
Rugby	1	4	11	13	22	30	35			
Tennis	3	7	10	21	27	36				
Other	8	14	16	19	20	25	26	39		
None	15	17	33	34	41	45	47	49	51	52

(a) Carry out a Kruskal–Wallis test, using the 5% significance level, to examine whether different sports achieved different results on average.

(b) Comment on your results in context.

(c) What assumption must be made for your results to be valid?

Key point summary

1 The **Mann–Whitney U test** is a **distribution-free**   *p196*
or **non-parametric test** which can be used to test for
differences between two sets of data which are **not
paired**.

2 All the data, in both samples, is **put in rank order**   *p200*
as though it was in one big group. The smallest value
is given rank value 1. The total of the rank values in
each sample, say, T_A and T_B, are found.

3 Two possible values exist for the test statistic, **U**, where: *p200*

$$\mathbf{U} = \mathbf{T} - \frac{n(n+1)}{2}$$

for T, the total of the rank values in the sample of size n.

4 Table 11 tabulates the **lower** tail critical values.   *p202*

5 In **two-tailed** tests, it is only necessary to compare   *p202*
the **lower** value of U with the **lower tail** critical value
from Table 11. If this **lower** test value is greater than
the critical value, then both possible U test values will
lie between the upper and lower critical values.

6 In one-tailed tests, $\mathbf{H_1}$, the alternative hypothesis, will   *p202*
determine which value of U should be compared with
the lower tail critical value. This will usually, but not
always, be the lower value of U.

7 The **Kruskal–Wallis test** is a **distribution-free** test   *p208*
which can be used to test for differences between more
than two populations.

8 All the data, in all samples, are put in **rank order** as   *p210*
though it was in one big group. The total of the rank
values in each of the samples, say, T_A, T_B and T_C, are found.

9 The test statistic is **H**, where: *p210*

$$\mathbf{H} = \frac{12}{N(N+1)} \sum_{i=1}^{k} \frac{T_i^2}{n_i} - 3(N+1).$$

T_i represents the sum of the ranks in sample i.
n_i represents the number of readings in sample i.
You do not need to memorise this formula. It is given in
the AQA formulae book.

10 The critical values are obtained from Table 6.   *p210*
The degrees of freedom are $k - 1$, where k is the number
of populations to be compared. The test is always
one-tailed.

11 The rejection of $\mathbf{H_0}$ leads to the conclusion that **at**   *p211*
least two of the populations differ.

9

Test yourself	What to review

1 For the following sets of data, say whether the data is paired or unpaired.

Section 9.1

  (a) Height (cm) of two groups of randomly selected males.

Office	146	163	178	154	179	180	142	155
Building site	173	167	182	158	168			

  (b) Pulse rate (beats per min) after 30 mins exercise for a group of 10 visitors to a gym, allocated to either leg exercise or weight training.

Leg exercise	194	186	220	180	130
Weight training	128	197	124	192	148

  (c) Times (seconds) to complete a task for seven people using first their right, then their left hand.

Person	A	B	C	D	E	F	G
Right	29	62	24	17	24	54	83
Left	44	42	29	27	18	76	94

2 Test the hypothesis that the average height of males who work in an office is lower than that of those who work on a building site using the data in **1(a)**.
Use the Mann–Whitney U test at the 5% significance level.

Section 9.2

3 A two-tailed Mann–Whitney U test is carried out on two samples, one of size 8 and the other of size 7. Obtain, using statistical tables, the upper tail and the lower tail critical values if a 5% significance level is used.

Section 9.3

4 For a two-tailed Mann–Whitney U test, using a 5% significance level, determine whether the following test statistics, **U**, where **U** is the lower of the two possible test values, would lead to the rejection of H_0.

Section 9.3

	n	m	U
(a)	8	12	15
(b)	11	10	49

| **Test yourself (continued)** | **What to review** |

5 The following data refers to the sex role attitude scores of male children from two separate groups.

Sections 9.2 and 9.3

Other males in family	No other males in family
23	17
21	18
9	19
30	16
17	14
22	10
19	11
20	

Carry out a Mann–Whitney U test, using a 5% significance level, to determine whether there is a difference on average between attitudes of male children with other males in family full-time and those with no other males in family.

6 For a Krustal–Wallis test carried out on the given numbers of samples, at the given significance level, state the critical value to be used.

Sections 9.4

	Number of samples	Significance level
(a)	3	5%
(b)	4	5%
(c)	5	5%
(d)	4	1%

7 Further research is done into the sex role attitude/ scores of male children, this time with three separate groups.

Sections 9.4

9

Father and male siblings of family	Absent father, male siblings in family	No other males in family
21	20	15
24	17	18
19	23	16
31	22	14

Carry out a Kruskal–Wallis test, using a 5% significance level, to determine whether there is a difference on average between attitudes of male children in the three separate groups.

Test yourself ANSWERS

1 (a) unpaired; **(b)** unpaired; **(c)** paired.

2 H_0 Population average height same for office and building workers
H_1 Population average height higher for building workers. One-tail 5%

O	2	6	10	3	11	12	1	4	$T_O = 49$	$m = 8$
B	9	7	13	5	8				$T_B = 42$	$n = 5$

Possible values $U = 49 - \left(\dfrac{8 \times 9}{2}\right) = 13$ or $U = 42 - \left(\dfrac{5 \times 6}{2}\right) = 27$, $cv = 8$.

For one-tail test, office workers are expected to be lower. $U = 13$.
$U > 8$, so accept H_0. Population height is the same.

3 $m = 8$, $n = 7$, two-tail 5%.
Lower tail $= 11$, upper tail $= (8 \times 7) - 11 = 45$.

4 (a) $m = 12$, $n = 8$, $U = 15$.
Two-tail 5%, lower $cv = 22$, $U < 22$. Reject H_0;
(b) $m = 11$, $n = 10$, $U = 49$.
Two-tail 5%, lower $cv = 27$, (upper $cv = 110 - 31 = 79$) $U > 27$, Accept H_0.

5 H_0 Population average attitudes same
H_1 Population averages differ. Two-tail 5%

Other males	14	12	15	1	13	$6\frac{1}{2}$	$9\frac{1}{2}$	11	$T_O = 82$	$m = 8$
No other males	$6\frac{1}{2}$	8	$9\frac{1}{2}$	5	4	2	3		$T_N = 38$	$n = 7$

Possible values $U = 82 - \left(\dfrac{8 \times 9}{2}\right) = 46$ or $U = 38 - \left(\dfrac{7 \times 8}{2}\right) = 10$.

For two-tail test, $U = 10$, $cv = 11$.
$U < cv$. Reject H_0. Population average attitudes differ.

6 (a) $cv = 5.99$; **(b)** $cv = 7.815$; **(c)** $cv = 9.488$ **(d)** $cv = 11.345$.

7 $H = 6.96$ $cv = 5.99$. (As there are only 4 in each group, cv may be unreliable.)
Significant evidence to doubt H_0 and conclude that at least two of the populations differ.
Significant evidence of a higher average score for attitude when father and male siblings in family than when no other males in family.

CHAPTER 10

Correlation

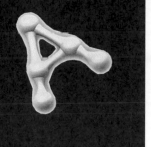

Learning objectives

After studying this chapter, you should be able to:
- test the hypothesis that a product moment correlation coefficient is equal to zero
- calculate Spearman's rank correlation coefficient for a sample
- test the hypothesis that the ranks of two variables are independent.

10.1 Testing a hypothesis about a product moment correlation coefficient

In S1, the concept of correlation was introduced and the method for evaluating a product moment correlation coefficient (PMCC) was given. Interpretation of a value for a sample PMCC was restricted to comments of the type 'very high positive correlation is evident' or 'slight negative correlation is evident'.

Carrying out a hypothesis test to determine whether a population PMCC, ρ, is **significantly** different from zero is very simple and involves a **test statistic** which is the sample PMCC, r, and its comparison with a **critical value** to be found in Table 8 of the formulae book.

> The **parameter**, ρ, is the population PMCC about which you will come to a conclusion after consideration of the sample **statistic**, r.

The simplest way to explain the procedure to follow when carrying out a hypothesis test on a population PMCC is to work through an example.

Consider the following data gathered on fuel consumption and speed:

Speed X (km per hour)	85	102	38	42	135	75	51
Fuel consumption Y (km per litre)	6.9	7.3	11.1	10.1	6.1	8.3	8.6

The PMCC for this sample data, $r = -0.913$, is found directly from a calculator.
The obvious comment is that there is very strong evidence to suggest an inverse linear relationship between speed and fuel consumption.

> If a different sample was chosen, a different value of r would have been obtained, but ρ, the population PMCC, remains constant.

To test this formally, the null and alternative hypotheses $\mathbf{H_0}$ and $\mathbf{H_1}$ must be stated at the start of the testing procedure.

$$\mathbf{H_0}\ \rho = 0$$

(**No** correlation exists in population between speed and fuel consumption.)

$$\mathbf{H_1}\ \rho < 0$$

(Population correlation coefficient is less than zero since we expect less kilometres per litre at higher speeds.)

This is a **one-tailed** test and we will use the standard **significance level** of 5%.

The **test statistic**, $r = -0.913$, has been calculated from the sample of seven pairs, $n = 7$.

The **critical value** relevant for this test is found from Table 8 by reading along the row where $n = 7$ and selecting the column which gives values for a one-tailed test at 5% significance level.

For these data, the **critical value** is -0.6694.

Note that the critical values given are all positive. The nature of the test being carried out will determine if a negative critical value is relevant.
The $\mathbf{H_1}$ indicates that we must consider extreme values for $\rho < 0$, that is extreme negative values.

The **test statistic** $r = -0.913$ is now compared with -0.6694.

If $r > -0.6694$, then we **accept $\mathbf{H_0}$** and conclude that there is **no significant** evidence that there is an inverse correlation between speed and consumption.

If, as in this case, $r \leqslant -0.6694$ we **reject $\mathbf{H_0}$** and conclude that there is **significant** evidence of correlation between speed and fuel consumption in the population.

The conclusion made is that the sample PMCC, $r = -0.913$, offers significant evidence, at the 5% level, of the existence of inverse correlation between speed and fuel consumption.

If $\rho = 0$, r is very unlikely to be exactly 0 but it is likely to be fairly close to 0.

Note that, in this test, the null hypothesis is that there is **no** correlation and this must be rejected for you to conclude that significant correlation exists in the population.

The test is **one-tailed** because the $\mathbf{H_1}$ involves $<$, that is ρ **less** than zero.

The **significance level** is the level of overwhelming evidence deemed necessary for the decision to conclude $\mathbf{H_0}$ is **not** true.

In this case, $\mathbf{H_0}$ is accepted if r is close to zero or above zero. The critical value defines what is meant by 'close to'.

If $r < -0.6694$ then $\mathbf{H_0}$ is rejected. The test is **one-tailed**.

The range of possible values for r which are equal to or less than the critical value is known as the **critical region**.

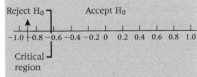

If a sample PMCC test statistic lies in the critical region, then $\mathbf{H_0}$ is rejected. If it does not lie in the critical region, then $\mathbf{H_0}$ is accepted.

There is a linear relationship between the speed and the fuel consumption in the population such that as speed increases, fuel consumption decreases.

The **critical values** given in Table 8 assume that the data comes from a **bivariate normal** distribution. Discussion of this distribution is beyond the scope of SS03.

Generally, it is safe to use these critical values unless some features of the data (such as an extreme outlier) suggests otherwise.

Note that the **critical values** given in Table 8 are for **one-tailed** or for **two-tailed** tests and care must be taken to ensure the correct **critical value** is used depending on the nature of the test concerned.

Worked example 10.1 involves another **one-tailed** test and Worked example 10.2 involves a two-tailed test.

> Conclusions should not just be given in general terms but should always refer to the specific data involved in the test.

> The **critical region** or **critical value** identifies the range of extreme values which lead to the rejection of H_0 $\rho = 0$.

Worked example 10.1

A group of antiques collectors is invited to estimate the likely price to be made at auction for a selection of eight different items. Their estimates, together with the actual prices achieved at auction, are given below.

Item	A	B	C	D	E	F	G	H
Price estimate	£350	£125	£75	£100	£25	£550	£750	£15
Actual price	£270	£140	£68	£85	£34	£390	£820	£18

Calculate the product moment correlation coefficient between the estimates made by the group of collectors and the actual prices achieved at auction.

Comment on the value of this correlation coefficient.

Stating clearly your null and alternative hypotheses, investigate whether there is a positive correlation between the estimated and actual prices. Use a 5% significance level.

> The question states that a positive correlation is to be tested for and so a **one-tailed** test is relevant.

Solution

The PMCC for this sample data, $r = 0.967$ is found directly from a calculator.

The **test statistic** $r = 0.967$.
The clear comment is that there is very strong evidence to suggest a positive correlation between estimated and actual price.

10

To test this formally, the null and alternative hypotheses H_0 and H_1 are stated.

$H_0 \, \rho = 0$

(**No** correlation exists in population between estimated and actual price.)

$H_1 \, \rho > 0$

(Population correlation coefficient is greater than zero.)

This is a **one-tailed** test with **significance level** of 5%.

The **test statistic**, $r = 0.967$, has been calculated from the sample of eight pairs, $n = 8$.

The **critical value** is found from Table 8 where $n = 8$ for a one-tailed test at 5%. The **critical value = 0.6215**.

In this case, $r > 0.6215$ and so there is significant evidence to reject H_0 and conclude that there is a strong positive correlation between estimated and actual prices.

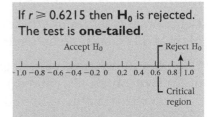

If $r \geqslant 0.6215$ then H_0 is rejected. The test is **one-tailed**.

Remember that if r is **less** than the **critical value** for this one-tailed test, we accept H_0 and conclude that there is no significant evidence of a correlation between prices.

Worked example 10.2

The results given show the yield, y, in grams from a chemical experiment corresponding to an input of x grams of a certain chemical to the process.

Input x gram	5.6	6.3	8.5	4.2	7.4	5.1	9.6	4.8	6.9	5.9
Yield y gram	82	78	86	65	91	80	75	72	89	74

Calculate the product moment correlation coefficient between the input and the yield.
Comment on the value of this correlation coefficient.
Stating clearly your null and alternative hypotheses, investigate whether there is an association between the input and the yield. Use a 5% significance level.

There is no suggestion that a positive or an inverse correlation is to be tested for and so a **two-tailed** test is relevant.

Solution

The PMCC for this sample data, $r = 0.484$ is found directly from a calculator.
The **test statistic,** $r = 0.484$.

This value for r indicates a slight, positive correlation between input and yield.

To test the significance of this result, the hypotheses are:

$H_0 \, \rho = 0$

(**No** correlation exists in population between input and yield.)

$\quad$ **H$_1$** $\rho \neq 0$

(Population correlation coefficient is different from zero.)

This is a **two-tailed** test with **significance level** of 5%.

The **test statistic**, $r = 0.484$ has been calculated from the sample of ten pairs, $n = 10$.

The **critical value** is found from Table 8 where $n = 10$ for a two-tailed test at 5%. The value in the table is 0.6319. As this is a two-tailed test, the critical values are ± 0.6319.

In this case, $|r| < 0.6319$ and so there is no significant evidence to doubt **H$_0$**.

We accept **H$_0$** and conclude that there is no significant evidence to indicate a correlation between input and yield in the population.

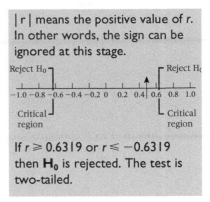

$|r|$ means the positive value of r. In other words, the sign can be ignored at this stage.

If $r \geqslant 0.6319$ or $r \leqslant -0.6319$ then **H$_0$** is rejected. The test is two-tailed.

EXERCISE 10A

1 A scientist, working in an agricultural research station, believes there is a relationship between the hardness of shells laid by chickens and the amount of a food supplement put into the diet of the chickens. He selects ten chickens of the same breed and gains the following results:

Chicken	Level of food supplement x (g)	Shell hardness y
A	7.0	1.2
B	9.8	2.1
C	11.6	3.4
D	17.5	6.1
E	7.6	1.3
F	8.2	1.7
G	12.4	3.4
H	17.5	6.2
I	9.5	2.1
J	19.5	7.1

(a) Calculate the product moment correlation coefficient between the level of supplement and the hardness of shell.

(b) Stating clearly your null and alternative hypotheses, investigate whether there is an association between level of supplement and shell hardness. Use a 5% significance level for this two-tailed test.

2 The body and heart mass of 14 10-month-old male mice are given below:

Body mass x (g)	Heart mass y (mg)
27	118
30	136
37	156
38	150
32	140
36	155
32	157
32	114
38	144
42	159
36	149
44	170
33	131
38	160

(a) Calculate the value of the product moment correlation coefficient for this data.

(b) Stating clearly your null and alternative hypotheses, test using a 5% significance level, whether there is a positive association between body mass and heart mass. This is a one-tailed test.

3 The following table gives the inflation rate, x, and the unemployment rate, y, for ten different countries during December 1979.

Country	Inflation rate x (%)	Unemployment rate y (%)
A	13.9	2.9
B	21.4	11.3
C	9.6	5.4
D	1.5	6.1
E	31.7	9.0
F	23.1	8.8
G	18.4	5.9
H	34.4	15.6
I	27.6	9.8
J	5.6	3.7

(a) Calculate the value of the product moment correlation coefficient between inflation and unemployment rates.

(b) Test, using a 5% significance level, the hypothesis $H_0 \rho = 0$ against the alternative hypothesis $H_1 \rho \neq 0$, where ρ is the population correlation coefficient between the rates.

4 In a workshop producing hand-made goods a score is assigned to each finished item on the basis of its quality (the better the quality the higher the score). The number of items produced by each of the 15 craftsmen on a particular day and their average quality score are given in the following table:

Craftsman	Items produced x	Average quality y
1	14	6.2
2	23	7.3
3	17	4.9
4	32	7.1
5	16	5.2
6	19	5.7
7	17	5.9
8	25	6.4
9	27	7.3
10	31	6.1
11	17	5.4
12	18	5.7
13	26	6.9
14	24	7.2
15	22	4.8

(a) Calculate the product moment correlation coefficient between number produced and quality score.

(b) Using a 5% significance level, test whether there is any evidence of an association between number produced and average quality score. State clearly your null and alternative hypotheses.

5 A hospital doctor was interested in the percentage of a certain drug absorbed by patients. She obtained the following data on 10 patients taking the drug on two separate days.

		Percentage of drug absorbed				
Patient		1	2	3	4	5
Day 1	x	35.5	16.6	13.6	42.5	28.5
Day 2	y	27.6	15.1	12.9	34.1	35.5
Patient		6	7	8	9	10
Day 1	x	30.3	8.7	21.5	16.4	32.3
Day 2	y	32.5	84.3	21.5	11.1	36.4

(a) Draw a scatter diagram of the data.

(b) Calculate the product moment correlation coefficient between the percentage of drug absorbed on Day 1 and Day 2.

(c) Stating clearly your null and alternative hypotheses, investigate whether there is an association between drug absorption on Day 1 and Day 2. Use a 5% significance level.

(d) After examining the scatter diagram, the doctor found that one of the points was surprising. Further checking revealed this point was the result of abnormal circumstances. The value of the product moment correlation coefficient for the remaining 9 points is 0.863.

(i) State which point has been omitted.

(ii) Test again, using a 5% significance level, whether there is an association between the drug absorption on the 2 days.   [A]

6 The following table shows the latitude, maximum and minimum temperature for 15 towns in a particular year.

Town	Latitude	Maximum temperature (°C)	Maximum temperature (°C)
A	60	8	0
B	53	14	6
C	52	14	9
D	44	18	6
E	60	10	2
F	45	17	8
G	50	12	6
H	60	8	1
I	39	20	14
J	56	11	5
K	40	22	11
L	53	13	4
M	49	16	7
N	52	12	10
O	42	25	12

(a) Calculate the product moment correlation coefficient between mid-temperature (mid-way between maximum and minimum temperature) and latitude.

(b) Investigate, using a 1% significance level, whether mid-temperature is associated with latitude.   [A]

Worked example 10.3

Two sets of times, measured in 1/100ths of a second, were obtained from nine subjects who were involved in two similar psychology experiments.

The product moment correlation coefficient for these nine pairs was found to be 0.879.

The hypothesis that there is no correlation between the results of the two experiments is to be tested using a 1% level of significance.

(a) Explain, in this context, what is meant by a Type I error.

(b) Explain, in this context, what is meant by a Type II error.

(c) Carry out the test.

Solution

(a) A Type I error is to conclude that there is a correlation between the results of the two experiments when, in reality, there is no correlation between the results in the population.

(b) A Type II error is to conclude that there is no correlation between the experimental results when, in reality, there is correlation between the experimental results in the population.

(c) $\mathbf{H_0} \; \rho = 0$
   $\mathbf{H_1} \; \rho \neq 0$
   The **critical value** for a two-sided 1% **significance level** with nine pairs of points is 0.7977.
   Since the **test statistic** is 0.879 $\mathbf{H_0}$ is rejected and we conclude, at the 1% significance level, that there is (positive) correlation between the experimental results in the population.

EXERCISE 10B

1 Refer to Question **1** in Exercise 10A and explain, in the context of this question, the meaning of a Type I error.

2 Refer to Question **2** in Exercise 10A and explain, in the context of this question, the meaning of a Type II error.

3 A local authority offers all its employees regular health checks. As part of the check, several physiological measurements are taken on each person. The results for two of the measurements X and Y on nine people are shown.

Person	X	Y
1	9	10
2	31	21
3	29	31
4	50	15
5	54	34
6	69	44
7	76	61
8	91	51
9	95	64

(a) Draw a scatter diagram of this data.

(b) Calculate the product moment correlation coefficient between the two measurements.

(c) Stating the null and alternative hypotheses used, test whether there is a positive association between the two measurements. The significance level is 5%.

10

(d) Explain in the context of this question the meaning of a Type I and a Type II error.

4 During the lambing season, eight ewes and their lambs were weighed at the time of birth with the following results:

Ewe	Weight of ewe x (kg)	Weight of lamb y (kg)
A	44	3.5
B	41	2.8
C	43	3.2
D	40	2.7
E	41	2.9
F	37	2.5
G	38	2.8
H	35	2.6

(a) Calculate the product moment correlation coefficient between weight of ewe and weight of lamb.

(b) Test whether these data could have come from a population with correlation coefficient $\rho = 0$.
Use a 1% significance level.

(c) Explain in the context of this question, the meaning of a Type I and a Type II error.

5 The following data is from the 1971 census, Northumberland County Report, Part III.

District	Population x (hundreds)	No. with no access to hot water supply y	No. in households with exclusive use flush toilets z (hundreds)
Bedlingtonshire	281	1225	240
Berwick	113	300	109
Blyth	343	1645	277
Gosforth	256	260	237
Hexham	92	160	81
Longbenton	486	680	454
Morpeth	128	145	120
Newbiggin	106	120	93
Newburn	392	795	359
Prudhoe	96	140	86
Seaton	318	620	284

Calculate the product moment correlation coefficient between:

(a) x and y,

(b) x and z,

(c) y and z.

(d) Stating the null and alternative hypotheses used, test, at a 5% significance level, whether there is an association between x and y, x and z or y and z.

(e) Discuss your results further in the light of the following comment from a housing expert: 'I would expect areas with a large number of people with no access to hot water supply to have a small number of people with exclusive use of inside flush toilets.' [A]

6 As part of an educational study, a random sample of ten boys from Harrowing School was given, after suitable preparation, three tests. These were an IQ test, a written test in Latin and a practical test in Music. Each boy was given a mark out of 100 on each test and these results are summarised in the following table:

Boy	Latin score (y)	IQ score (x)	Music score (z)
A	58	73	78
B	67	84	50
C	41	55	45
D	59	69	42
E	61	75	58
F	45	52	35
G	78	79	55
H	53	72	52
I	50	65	38
J	85	89	62

(a) Calculate the product moment correlation coefficient between:

   (i) IQ score and Latin score,

   (ii) IQ score and Music score,

(b) Stating the null and alternative hypotheses, test, at the 5% level of significance, whether there is a positive association between IQ and Latin scores or between IQ and Music scores.

(c) Would your answers to (b) be affected if the tests had been carried out at the 1% significance level? [A]

10.2 Spearman's rank correlation coefficient, r_s

Consider the following example which will introduce rank correlation.

Worked example 10.4

At a film festival in the South of France, two experienced judges are each asked to view seven films in the 'Comedy' category and put them in rank order where a rank of 1 is given to the best of the seven films and a rank of 7 is given to the one judged to be the worst.

The following results were obtained:

Film	1	2	3	4	5	6	7
Judge A(x)	7	3	1	4	2	5	6
Judge B(y)	1	2	6	7	4	3	5

Find the value of the correlation coefficient between the rankings of the two judges and comment on your result.

Solution

The data given here differ from that in Section 10.1 where the bivariate data given were measured variables or scores **not** *ranks* as is the case in this example.

However, the value of the correlation coefficient between the ranks of the two judges can be found in the same way as in S1 chapter 7.

> As in Worked examples 10.1 and 10.2, the value of r can be found directly from a calculator and this is the easiest way to evaluate it.

$$r = \frac{S_{XY}}{\sqrt{S_{XX} \times S_{YY}}}$$

$$S_{XY} = 100 - \frac{28^2}{7} = -12$$

$$S_{XX} = 140 - \frac{28^2}{7} = 28$$

and

$$S_{YY} = 140 - \frac{28^2}{7} = 28$$

> Note that S_{XX} and S_{YY} are the same, both equal to 28, because the x and y values are the same seven rank values 1 to 7.

So $$r = \frac{-12}{\sqrt{28 \times 28}} = \frac{-3}{7} = -0.42857 = -0.429 \text{ (3 s.f.)}$$

The correlation coefficient between the ranks is called **Spearman's rank correlation coefficient** and is usually denoted r_s. In this case it indicates that the two judges do not seem to agree in their opinions of the films since it is **negative.** It would appear that Judge A thinks highly of some films which Judge B does not like very much at all.

> Just as with the product moment correlation coefficient, this measure of correlation always lies in the range:
>
> $$-1 \leqslant r_s \leqslant +1$$

> Spearman's rank correlation coefficient, r_s, provides a measure of the association between the **rank orders** of two variables.

10.3 An alternative formula for r_s

A quite different formula is given in the formulae book for the Spearman's rank correlation coefficient, r_s. This formula is equivalent to the calculation above, but uses the sums of simple series, since rank values are involved, to simplify the formula to

$$r_s = 1 - \frac{6\sum d^2}{n(n^2 - 1)},$$

where d represents the difference between the rank values and n is the sample size.

Film	1	2	3	4	5	6	7
Judge A(x)	7	3	1	4	2	5	6
Judge B(y)	1	2	6	7	4	3	5
$|d|$		6	1	5	3	2	2	1

Since the values for d are all squared, just the positive values for d, or $|d|$, are given in the table opposite.

$\sum d^2 = 6^2 + 1^2 + 5^2 + 3^2 + 2^2 + 2^2 + 1^2 = 80$ and $n = 7$

So $r_s = 1 - \dfrac{6 \times 80}{7 \times 48} = \dfrac{-3}{7} = -0.429$ (3 s.f.) as before.

In practice, finding r_s directly from a calculator, using the rank values, is easier than using this formula and is the recommended method.

Spearman's rank correlation coefficient can be obtained from the **rank values** of data, either by using the formula

$$r_s = 1 - \frac{6\sum d^2}{n(n^2 - 1)}$$

or by obtaining the value of the product moment correlation coefficient between these rank values directly from the calculator.

If the data contains ties the two methods give slightly different results. The PMCC between rank values is correct but either value will be accepted in an examination.

Worked example 10.5

Eight contestants enter a writing competition for short stories. Two judges place these contestants in the following order where a rank value of 1 is given to the best story and a rank value of 8 to the one least liked by the judge.

The results are in the table below:

Contestant	A	B	C	D	E	F	G	H
Judge 1(x)	8	6	7	2	4	3	5	1
Judge 2(y)	5	2	4	8	1	6	7	3

Calculate the coefficient of rank correlation and comment on your result.

Solution

Using the formula for Spearman's rank correlation coefficient:

$$r_s = 1 - \frac{6\sum d^2}{n(n^2 - 1)}$$

Alternatively the rank values can be entered into your calculator and r_s obtained directly. You are recommended to do this in the exam.

Contestant	A	B	C	D	E	F	G	H
Judge 1(x)	8	6	7	2	4	3	5	1
Judge 2(y)	5	2	4	8	1	6	7	3
$|d|$	3	4	3	6	3	3	2	2

10

$$n = 8 \qquad \sum d^2 = 3^2 + 4^2 + 3^2 + 6^2 + 3^2 + 3^2 + 2^2 + 2^2 = 96$$

$$\text{So } r_s = 1 - \frac{6 \times 96}{8 \times 63} = -\frac{1}{7} = -0.143$$

This value for the rank correlation coefficient is very close to zero and indicates little connection between the rank values assigned to each story by the two judges. They appear to have very different opinions about which stories are the best.

Worked example 10.6

Data is gathered on the number of television licences per 1000 of population and the level of cinema admissions per head per year in ten large towns in the North of the UK. The figures are given in the following table:

Town	A	B	C	D	E	F	G	H	J	K
TV licences per 1000	220	80	160	290	395	75	300	325	440	340
Number of cinema admissions per head	10.8	12.1	12.5	12.2	8.8	13.5	10.0	9.5	9.2	10.5

(a) Explain the differences and the similarities between the product moment correlation coefficient and Spearman's rank correlation coefficient.

(b) Calculate the value of Spearman's rank correlation coefficient for the above data.

Solution

(a) For the product moment correlation coefficent, the paired data must be scores or measured items. For Spearman's rank coefficient, the data involved must be rank order values. The product moment correlation coefficient measures how close the given bivariate data is to a straight line relationship. Spearman's rank coefficient quantifies how closely connected the bivariate data is with respect to rank order. Both coefficients can range from -1 to $+1$.

(b) The data given is not rank order values and therefore the given data has to be replaced with its rank order values to enable Spearman's rank correlation coefficient to be calculated. In the table below, the rank values are substituted for the original given measured values.

Town	A	B	C	D	E	F	G	H	J	K
Rank TV licences	7	9	8	6	2	10	5	4	1	3
Rank no. of cinema admissions	5	4	2	3	10	1	7	8	9	6
$|d|$	2	5	6	3	8	9	2	4	8	3

$n = 8$ $\sum d^2 = 4 + 25 + 36 + 9 + 64 + 81 + 4 + 16 + 64 + 9 = 312$

So $r_s = 1 - \dfrac{6 \times 312}{10 \times 99} = -\dfrac{49}{55} = -0.891$

The value of r_s is fairly close to -1 and therefore this rank coefficient indicates a close inverse connection in the rank order values for TV licences and cinema admissions. It would seem that towns where there are high numbers of TV licences per head tend to be those where there are lower cinema admissions.

> Again, by entering the rank values into your calculator you can obtain the value for r_s directly.

Worked example 10.7

A wine taster is given nine bottles of wine to try in a blind tasting experiment. She is asked to place these wines in order of preference (rank 1 for the best) and these results are shown below, together with the price of each bottle.

Bottle	A	B	C	D	E	F	G	H	J
Price (p)	295	345	395	445	595	595	650	775	950
Rank	8	7	4	9	1	6	2	3	5

Calculate the value of Spearman's rank correlation coefficient between the rank order value for the wine and its price. Comment on the value of this correlation coefficient.

Solution

The preferences of the taster are already ranked but the prices of the bottles are given as their actual values, not rank values, and therefore these need to be put in rank order.

Note that two bottles of wine are the same price and the usual convention for tied ranks is followed.

Bottle	A	B	C	D	E	F	G	H	J
Price (p)	9	8	7	6	4.5	4.5	3	2	1
Rank	8	7	4	9	1	6	2	3	5
$|d|$	1	1	3	3	3.5	1.5	1	1	4

> The convention for tied ranks is that the *average* rank value, for those which tie, should be assigned to each reading. In this case, ranks 4 and 5 tie so each value is assigned the rank 4.5.

$n = 9$ $\sum d^2 = 1 + 1 + 9 + 9 + 12.25 + 2.25 + 1 + 1 + 16 = 52.5$

So $r_s = 1 - \dfrac{6 \times 52.5}{9 \times 80} = \dfrac{9}{16} = 0.5625$

This value for r_s indicates a fair, positive connection between the rank order values for preference and for price. It appears that those wines preferred most by the taster often tended to be the more expensive ones. However, this rule was not always the case and some of the cheaper wines were quite highly rated.

> If you directly put these rank values into the calculator, the value for r_s is slightly different because of the tied ranks. The calculator value is $r_s = 0.561$. This is the correct value, but **either** value is acceptable in the exam.

10

EXERCISE 10C

1 A child was asked to rank seven types of sweet according to preference and to sweetness with the following results:

Type	1	2	3	4	5	6	7
Preference	3	4	1	2	6	5	7
Sweetness	2	3	4	1	5	6	7

Calculate Spearman's rank correlation coefficient for these data. Comment on your result.

2 A food critic tastes eight supermarket instant ready meals and puts the meals in rank order of preference where rank 1 indicates the meal he considered to be the best.
The results, together with the prices of the meals are given in the table below:

Brand	A	B	C	D	E	F	G	H
Rank preference	5	4	6	7	8	1	2	3
Price (p)	171	165	127	150	143	139	149	179

Calculate the value of Spearman's rank correlation coefficient between the rank of preference and the price of the meals.

3 Nine chrysanthemums were ranked by a judge according to their colour and their overall appearance. The results are shown below:

Flower	1	2	3	4	5	6	7	8	9
Colour	2	1	9	4	8	7	6	5	3
Overall	6	5	7	2	4	9	1	3	8

Find the value of the rank correlation coefficient between the colour and the overall appearance of the flowers.

4 The table below gives the number of rotten peaches in 10 randomly selected crates from a large consignment after they had been kept in storage for the stated number of days.

Days in storage	6	6	14	21	24	29	30	35	38	40
Number of rotten peaches	3	8	8	6	15	11	13	17	18	21

Calculate Spearman's rank correlation coefficient for these data and comment on your findings.

5 A group of students attempt an assignment consisting of two questions. The scores on each question for a random sample of eight of this group are given below.

Student	1	2	3	4	5	6	7	8
Question 1	42	68	32	84	71	55	55	70
Question 2	39	75	43	79	83	65	62	68

(a) Calculate the Spearman's rank correlation coefficient between the scores on the two questions.

(b) Give an interpretation of your result.

6 A group of students were assessed by written examination, practical work and essay. The following table shows the examination percentage, practical work grade and the essay rank (best essay ranked 1) as judged by the course tutor.

Student	G	H	I	J	K	L	M	N
Exam %	62	43	57	84	41	17	29	66
Practical grade	A−	C	C+	A	C−	B+	D	B+
Essay rank	2	6	4	1	7	5	8	3

(a) Calculate Spearman's rank correlation coefficient between:
   (i) exam percentage and practical grade,
   (ii) exam percentage and essay rank,
   (iii) practical grade and essay rank.

(b) Comment on the performance of student L.

(c) It is suggested that next year only examination mark and practical work grade should be recorded. Comment on this.

(d) Why is Spearman's rank more suitable than the product moment correlation coefficient for these data? [A]

10.4 Testing a hypothesis about Spearman's rank correlation coefficient

In section 10.1 of this chapter, the procedure for testing whether a population product moment correlation coefficient is significantly different from zero was introduced. In exactly the same way, a test can be carried out to determine whether, in a population, the rank orders of two variables are independent. Note that it is not possible to rank all the members of a population model and so Spearman's rank correlation coefficient for a population does not exist.

> The critical values for testing Spearman's rank correlation coefficient are different than for the PMCC. This is because the PMCC values assume that the data follows a bivariate normal distribution. This will not be true if the data consists of ranks.

10

The critical values for testing a Spearman's rank correlation coefficient are found in Table 9 of the formulae book. On comparison, it can be seen that the values given are slightly lower than those in Table 8 which relate to values of the product moment correlation coefficient, although for large samples there is hardly any difference.

In the same way as a product moment correlation coefficient is tested, the sample rank correlation coefficient is compared with the critical value obtained from Table 9 for the appropriate significance level and the sample size n.

> A test can be carried out to determine whether an association between rank orders of two variables exists in the population. The null hypothesis:
>
> H_0 ranks, in population, are independent
>
> The critical value is found from Table 9.

Because r_s is calculated from ranks it can only take certain values (it is a discrete variable) and so it is not usually possible to obtain critical values with exactly the required significance level. The critical values given, in Table 9, are those with significance levels **closest** to the value stated. This means that the value tabulated for 5% may give a significance level which is not exactly 5% but is a little less or a little more.

In Table 8, the critical values relate to the exact significance level since r is a **continuous** variable.

In practice, the critical values found in Table 9 will be used in exactly the same way as they were when found using Table 8.

Worked example 10.8

In a major art competition, two judges award scores out of a total of 40 to the eight finalists. The scores given to these eight finalists are:

Painting	1	2	3	4	5	6	7	8
Judge A	39	23	25	35	37	36	26	30
Judge B	36	16	28	30	31	24	19	20

Calculate the value of Spearman's rank correlation coefficient between the scores given by the two judges.
Stating clearly the null and alternative hypotheses used, test whether there is any evidence of an association between the two judgements.

H_0 and H_1 are expressed differently to the product moment correlation coefficient. This is because a population value of Spearman's correlation coefficient does not exist.

Solution

The rank values for the above data are given below:

Painting	1	2	3	4	5	6	7	8
Judge A	1	8	7	4	2	3	6	5
Judge B	1	8	4	3	2	5	7	6

Directly from entering these ranks into the calculator, it can be found that $r_s = 0.810$.

H_0 pop ranks are independent
H_1 pop ranks are not independent

From Table 9, the critical value for a sample where $n = 8$, two-tailed at 5% is ± 0.7381.
The test statistic is $r_s = 0.810$.

As before, it is recommended that a calculator is used to obtain r_s.

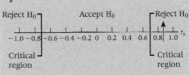

$r_s \geqslant 0.7381$ or $r_s \leqslant -0.7381$ then H_0 is rejected. The test is two-tailed.

In this case, $|r_s| > 0.7381$ and so there is significant evidence to reject H_0 and conclude that there is an association in the rank order of the scores between the two judges. The paintings to which Judge A awarded higher scores were also those awarded higher scores by Judge B. The two judges seem to agree fairly well in their opinions of the paintings.

Worked example 10.9

Two children were asked to score 10 chocolate bars according to their sweetness. The scores were on a scale of 1 to 10 where 1 is the least sweet and 10 the sweetest. One of the children scored the bars in the reverse order by mistake.

The results are shown below:

Bar	Child 1	Child 2
A	3	5
B	4	10
C	1	9
D	6	6
E	2	8
F	7	1
G	5	7
H	7	4
I	9	3
K	8	2

(a) Calculate the Spearman's rank correlation coefficient between the scores of the two children.

(b) Stating clearly your null and alternative hypotheses, test, at the 5% level, whether there is an inverse association between the rank values of the sweetness scores given by the two children.

Solution

The original data is not given as rank values and so the data must be put in rank order. The ranks are given in the table below. A rank value of 1 is assigned to the highest score (the sweetest) and a rank value of 10 to the least sweet.

Bar	Child 1	Child 2	d
A	8	6	2
B	7	1	6
C	10	2	8
D	5	5	0
E	9	3	6
F	3.5	10	6.5
G	6	4	2
H	3.5	7	3.5
I	1	8	7
J	2	9	7

Note here that there are tied ranks. Child 1 gave bars F and H the same score of 7. The rank values 3 and 4 are therefore averaged and a rank of 3.5 given to each.

10

(a) Directly from the calculator, the value for r_s can be obtained. In this case, $r_s = -0.802$.

Using the formula, $\sum d^2 = 296.5$ and so,

$$r_s = 1 - \frac{(6 \times 296.5)}{(10 \times 99)} = -0.797$$

(b) H_0 population ranks are independent

H_1 inverse association between population ranks

The critical value from Table 9 for a one-tailed test using a 5% significance level for a sample where $n = 10$ is -0.5636. In this case $r_s = -0.802$ (or -0.797) and $r_s < -0.5636$.

We conclude that H_0 is rejected and there is evidence of an inverse association between the rank order values of sweetness given to the chocolate bars by the two children.
The bars which child 1 scored higher for sweetness, child 2 tended to score lower.

In the case of tied ranks, the formula gives a slightly different result to that obtained directly from the calculator. This does not affect the conclusion of the hypothesis test. Both methods are given in the example and both are equally acceptable in the exam. However the value obtained directly from the calculator is the correct one.

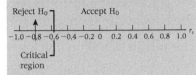

If $r_s \leqslant -0.5636$ then H_0 is rejected. The test is one-tailed.

EXERCISE 10D

1 Stating the null and alternative hypotheses you are using, test the following Spearman's rank correlation coefficients to determine whether there is:

(a) an association between 'preference for' and 'price of' ready meals as given in Question **2** of Exercise 10C. Use a 5% significance level;

(b) a positive association between preference and sweetness using the data given in Question **1** of Exercise 10C. Use a 1% significance level;

(c) a positive association between colour and overall appearance rating using the data given in Question **3** of Exercise 10C. Use a 1% significance level;

(d) an association between number of days in storage and number of rotten peaches using the data of Question **4** of Exercise 10C. Use a 1% significance level.

2 The five finalists in a piano competition were placed in the following order by the two judges:

	1st	2nd	3rd	4th	5th
Judge 1	C	E	D	A	B
Judge 2	B	C	A	D	E

Evaluate the Spearman's rank correlation coefficient for these data.

Stating the null and alternative hypothesis used, test, at the 5% significance level, for an inverse association between the opinions of the judges and comment.

3 The following table shows the percentage of part-time staff employed by nine supermarkets in a large city and their rankings in terms of weekly takings.

Part-time staff, %	45	56	38	29	54	33	36	33	22
Weekly takings, rank	7	4	5	9	2	8	1	3	6

Calculate the value of Spearman's rank correlation coefficient for this set of data. Test, at the 5% significance level, the hypothesis that there is no relationship between the percentage of part-time staff employed and weekly takings.

4 The reading age of a random sample of eight children entering secondary school was recorded. During their first weeks their English teacher asked them to write a poem. The reading ages and the ranks of the poems, as judged by the English teacher are shown in the following table.

Reading age	8.7	11.2	14.4	6.8	12.3	13.1	9.6	10.8
Rank of poem	7	4	3	8	1	6	5	2

Calculate Spearman's rank correlation coefficient and test, at the 1% significance level, for an association between reading age and ability to write poetry.

5 Rainfall, x cm, and hours of sunshine, y, on nine randomly selected October days is shown in the table below.

x	1.3	3.8	4.2	2.6	2.1	2.6	5.3	0.0	0.9
y	1.5	0.3	0.0	4.2	3.6	0.5	0.0	6.2	1.4

Calculate Spearman's rank correlation coefficient. Investigate, at the 5% significance level, whether days with high rainfall tend to have few hours of sunshine.

10.5 The relationship between product moment and Spearman's rank correlation coefficients

In some cases, the values of r and r_s are very similar and both will lead to the conclusion that there is an association between the two variables in the population. In this case, a scatter diagram will show a clear linear relationship.

As you saw in S1, Chapter 7, it is always useful to refer to a scatter diagram when interpreting correlation coefficients.

10

However, sometimes the value of $|r_s|$ is much greater than the value of $|r|$. This is a clear indication that there is a non-linear relationship between the two variables which can be seen by referring to a scatter diagram.

Consider the following data which gives test results for nine students selected at random from Year 8 at a large high school. The table also shows the rank values in brackets.

Student	Maths	English	Physics
1	9 (9)	59 (1)	4 (9)
2	41 (6)	38 (4)	17 (6)
3	49 (5)	39 (3)	32 (5)
4	18 (8)	51 (2)	11 (8)
5	52 (4)	31 (6)	43 (4)
6	63 (1)	23 (9)	74 (2)
7	62 (2)	30 (7)	83 (1)
8	32 (7)	36 (5)	14 (7)
9	58 (3)	24 (8)	55 (3)

> Values of the correlation coefficients are obtained directly from a calculator.

The scatter diagram illustrates the students' results in maths and English.
The value of the product moment correlation coefficient is $r = -0.941$.
The value of the Spearman's rank correlation coefficient is $r_s = -0.917$.
Both coefficients are fairly similar and indicate a strong inverse relationship between the results in maths and English for Year 8 students.

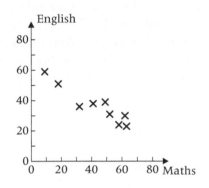

The second scatter diagram is for results in maths and in physics.
The value of the product moment correlation coefficient is $r = 0.894$.
The value of the Spearman's rank correlation coefficient is $r_s = 0.983$.
$|r_s|$ is much larger than $|r|$ and the scatter diagram reveals the non-linear relationship.

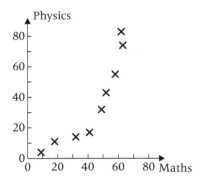

MIXED EXERCISE

1 A clothing manufacturer collected the following data on the maintenance costs and age of his sewing machines.

Sewing machine	Maintenance cost (£)	Age (months)
1	24	13
2	144	75
3	110	64
4	63	52
5	240	90
6	20	15
7	40	35
8	180	82
9	42	25
10	50	46
11	92	50

(a) Plot a scatter diagram of the data.

(b) Calculate the product moment correlation coefficient.

(c) Calculate the Spearman's rank correlation coefficient.

(d) Stating clearly the null and alternative hypotheses used, carry out two tests, using your results from **(b)** and **(c)**, to determine whether there appears to be an association between age and costs of sewing machines. Use a 1% significance level for each test.

2 The following data show the IQ and the score in an English test of a random sample of 10 pupils taken from a mixed ability Year group.

Pupil	IQ	English (%)
A	110	52
B	107	62
C	127	74
D	100	40
E	132	70
F	130	68
G	98	46
H	109	76
I	114	62
J	124	72

(a) Find the product moment correlation coefficient.

(b) Stating clearly the null and alternative hypotheses used, test whether there is a direct association between IQ and English score. Use a 5% significance level.

For another group of eight randomly selected pupils from the mixed ability Year group, their teacher assessed each pupil according to their aptitude for schoolwork and their perseverance. A rating scale of 0 to 100 was used.

Pupil	Aptitude	Perseverance
K	42	73
L	68	39
M	32	63
N	84	49
O	71	83
P	55	65
Q	58	62
R	70	68

(c) Find the Spearman's rank correlation coefficient for these data.

(d) Stating clearly the null and alternative hypotheses used, test whether there is an association between aptitude and perseverance. Use a 5% significance level.

(e) Comment on your findings in **(b)** and **(c)**.   [A]

10

3 A company employing sales representatives wishes to assess the correlation between their scores obtained during training and during their first few years of service.

During training, points are awarded: these reflect efficiency at work, punctuality, appearance, etc. Once working for the company, points are awarded for the amount of business brought in, number of contacts made, customer responses, etc. Total scores are found from these points.

For a random sample of 12 sales representatives, the training and service scores awarded by one manager in the company are given in the following table.

Sales representative	Training score	Service score
A	10	11
B	15	11
C	20	27
D	19	25
E	9	11
F	12	13
G	11	14
H	7	9
I	18	15
J	21	26
K	12	20
L	17	18

(a) Calculate the Spearman's rank correlation coefficient for the data.

(b) Stating the null and alternative hypotheses used, test for a positive association between training and service scores awarded by this manager. Use a 5% significance level. Interpret your findings in the context of this question.

(c) A second manager also assessed the same 12 sales representatives. This manager agreed with the service scores awarded but awarded very different training scores. The Spearman's rank correlation coefficient resulting from the scores awarded by the second manager was −0.808.
Carry out a test using a 5% significance level to determine whether there appears to be an association between the scores awarded by the second manager. Comment on your findings. [A]

4 The following data show the annual income per head, x ($) and the infant mortality, y (per 1000 live births) for a sample of 11 countries.

Country	x	y
A	130	150
B	5950	43
C	560	121
D	2010	53
E	1870	41
F	170	169
G	390	143
H	580	59
I	820	75
J	6620	20
K	3800	39

(a) Draw a scatter diagram for the data and describe the relationship between income and infant mortality as suggested by the scatter diagram.

(b) An economist asks you to calculate the product moment correlation coefficient.

   (i) Carry out this calculation.

   (ii) Explain briefly to the economist why this calculation may not be appropriate.

(c) Calculate Spearman's rank correlation coefficient for this data.

(d) Stating clearly the null and alternative hypotheses used, test to see whether there is an association in rank orders between income and infant mortality. Use a 1% significance level.

(e) Comment and compare the values of the two correlation coefficients you have calculated. [A]

5   A consumer company tests 10 microwave ovens and awards a grade according to their efficiency when used for various tasks. The grade awarded for efficiency is given in the following table, together with the price and the rank order value given for the overall appearance of each oven.

Oven	Efficiency grade	Price (£)	Appearance rank
1	A	256	3
2	C	149	6
3	D	150	9
4	B+	199	4
5	C	175	7
6	E	142	10
7	D	177	8
8	B+	185	1
9	B	190	5
10	A+	275	2

(a) Calculate Spearman's rank correlation coefficient between:
   (i) price and efficiency grade,
   (ii) price and appearance ranking.

(b) Investigate, at the 1% significance level, whether the price and the efficiency grades are related. State clearly the null and alternative hypotheses used.

(c) It is suggested that a wider range of models of oven should be investigated but that appearance ranking should not be used. Comment on this suggestion with reference to your results in **(a)** and **(b)**.

(d) Comment also on the suggestion that the product moment correlation coefficient between price and efficiency grade should be calculated.   [A]

6 The length of service and the gross earnings of a sample of 11 employees of a large store in 1984, are given below:

Employee	A	B	C	D	E	F	G	H	J	K	L
Length of service (months)	132	64	28	117	94	14	17	19	76	21	18
Gross earnings 1984 (£)	8220	6140	4890	7410	5930	4720	13 680	4320	6490	5070	4760

(a) Plot a scatter diagram of these data.

(b) Calculate the product moment correlation coefficient.

(c) Calculate Spearman's rank correlation coefficient.

(d) Use your result in **(c)** to test, at the 5% significance level, whether gross annual earnings are associated with length of service.

(e) Describe the relationship (if any) between length of service and gross earnings suggested by the scatter diagram and comment on the values of the two correlation coefficients calculated.   [A]

7 An instrument panel is being designed to control a complex industrial process. It will be necessary to use both hands independently to operate the panel. To help with the design it was decided to time a number of operators, each carrying out the same task once with the left hand and once with the right hand.

The times, in seconds, were as follows:

Operator	A	B	C	D	E	F	G	H	I	J	K
l.h. x	49	58	63	42	27	55	39	33	72	66	50
r.h. y	34	37	49	27	49	40	66	21	64	42	37

(a) Plot a scatter diagram of the data.

(b) Calculate the product moment correlation coefficient between the two variables and comment on this value.

(c) Test, at the 5% significance level, for an association between times with left hand and times with right hand.

(d) Further investigation revealed that two of the operators were left-handed. State, giving a reason, which you think these were. Omitting their two results, calculate Spearman's rank correlation coefficient and test for an association.

(e) What can you say about the relationship between the times to carry out the task with left and right hands?

[A]

8 An archaeologist has access to two scientific processes for determining the age of ceramic fragments found in excavation of ancient sites. She wishes to investigate whether the two processes produce similar results. The results for the ages, in hundreds of years, of ten fragments are as follows:

Fragment	A	B	C	D	E	F	G	H	I	J
Age Process 1	152	123	105	145	107	128	115	143	135	128
Age Process 2	155	121	107	144	105	126	112	144	139	129

(a) Plot a scatter diagram for the above data and state what this indicates regarding the association between the ages given by the two processes.

(b) Calculate the value of the product moment correlation coefficient between the ages given by process 1 and process 2.

(c) Stating the null and alternative hypotheses used, carry out a test at the 1% level of significance to investigate whether your calculated value in (b) indicates a direct association between the ages given by the two processes.

A colleague of this archaeologist claims that he can determine the rank order of the ages of ceramic fragments without the use of any scientific processes. The rank values he assigned to the ten fragments, where rank value 1 is the oldest, are given in the following table.

Fragment	A	B	C	D	E	F	G	H	I	J
Age	8	3	5	9	2	7	4	1	10	6

(d) Calculate the value of Spearman's rank correlation coefficient between the ages given by **process 1** and the ages determined by the colleague.

(e) Carry out a hypothesis test, at the 5% level of significance, to investigate whether your calculated value in (d) indicates an association between the two sets of ages. Your conclusion should refer to the claim made by the colleague.

[A]

10

9 A group of students planned to share a house whilst at university. They identified 12 possible houses and used a map to measure the straight line distance, x km, of each house from the Student Union. They then measured the road distance, y km, of each house from the Student Union. The distances are shown below.

House	A	B	C	D	E	F	G	H	I	J	K	L
Straight line distance, x	7.5	3.0	23.5	13.4	9.3	9.1	9.8	3.7	17.9	4.7	2.0	2.4
Road distance, y	8.8	3.2	28.4	16.7	9.5	8.9	12.4	9.9	22.5	4.9	2.5	2.9

(a) Draw a scatter diagram of the data.

(b) A statistician recommended that houses F and H should be omitted from any further analysis as she suspected errors in their data. Discuss why she made this recommendation and, for each house, the strength of the evidence supporting her suspicions.

(c) The houses in the table, have been presented in order of preference. That is, the students would prefer to live in house A, followed by B, etc. Omitting houses F and H calculate the value of Spearman's rank correlation coefficient between preference and road distance from the Student Union. Investigate, using a 5% significance level, whether the ranks are related. [A]

10 Building Societies are assessed annually in terms of customer satisfaction ratings. For the year 2000, the assets (in £ million) and the customer satisfaction rating (out of a maximum of 20), for ten Building Societies, are given in the following table:

Building Society	Assets	Customer satisfaction rating
A	15 789	15
B	59 810	13
C	17 831	14
D	11 708	19
E	13 040	17
F	9824	18
G	21 265	17
H	35 012	10
I	21 499	12
J	8242	16

(a) Find the value of Spearman's rank correlation coefficient between customer satisfaction rating and assets.

(b) Carry out a hypothesis test, at a 5% level of significance, to determine whether your value calculated in **(a)** indicates an association between customer rating and assets. Interpret your conclusion in the context of this question.

(c) The number of repossessions made by Building Societies is also recorded. For the year 2000, the figures for the same ten Building Societies are given in the following table:

Building Society	Number of repossessions
A	68
B	226
C	89
D	37
E	46
F	22
G	98
H	146
I	114
J	12

(i) Find the value of the product moment correlation coefficient between assets and number of repossessions.

(ii) Carry out a hypothesis test, at the 1% significance level, to determine whether there is a positive association between assets and number of repossessions. [A]

11 In a study of electronic calculators, an index was devised to measure the variety and complexity of the operations they were capable of performing. The value of the index for 10 different calculators together with their selling price is given below.

(a) Plot a scatter diagram of the data.

(b) Calculate Spearman's rank correlation coefficient.

(c) Give a reason why it is more appropriate to calculate Spearman's rank rather than the product moment correlation coefficient for this set of data.

Selling price, £x	10	40	60	92	109	120	128	160	164	170
Index, y	15	38	27	70	98	170	162	305	485	599

(d) Stating clearly your null and alternative hypotheses, investigate each of the following claims:
  (i) selling price is associated with the value of the index; use a 10% significance level;
  (ii) high selling price is associated with high values of the index; use a 1% significance level;

      (iii) high selling price is associated with low values of
          the index; use a 5% significance level.

   (e) Explain why it was unnecessary to identify the critical
      value in **(d)(iii)**.   [A]

Key point summary

1 The **critical value** is found directly from statistical   *p223*
tables (Table 8) for testing a PMCC. $\mathbf{H_0}\ \rho = 0$.

2 Spearman's rank correlation coefficient, r_s, provides a   *p232*
measure of the association between the rank orders of
two variables. If the data is not given as rank order
values, then it must be put into rank order before
calculations are carried out.

3 Spearman's rank correlation coefficient can be   *p233*
obtained from the **rank values** of data, either by
using the formula

$$r_s = 1 - \frac{6\sum d^2}{n(n^2 - 1)},$$

where d is the difference in rank order values for each pair
and n is the number of pairs in the sample; or by obtaining
the value of the product moment correlation coefficient
between these rank values directly from the calculator.
This is the **recommended** method.

If the data contains ties the two methods give slightly
different results. The PMCC between rank values is
correct but either value will be accepted in an examination.

4 A test can be carried out to determine whether an   *p238*
association between rank orders of two variables
exists in the population. The null hypothesis is

    $\mathbf{H_0}$ ranks in population are independent.

The critical value is found from Table 9.

Test yourself   **What to review**

1 A manufacturer collects data on the age and the annual   *Sections 10.1*
maintenance costs for his 12 welding machines.
The product moment correlation coefficient between age and
cost is found to be 0.874.
Stating clearly your null and alternative hypotheses and
using a 5% significance level, test whether there is any
evidence of a positive association between age and
maintenance costs for these machines.

2 At the end of their first year on a college course, students take several exams, two of which are sociology and research methods. The product moment correlation coefficient is evaluated between the sociology and the research methods scores for a randomly chosen selection of 15 students. This coefficient was evaluated as -0.675.

Sections 10.1

(a) State clearly the null and alternative hypotheses to be used in testing for an inverse association between the scores of students in sociology and in research methods.

(b) Carry out this test using a 1% significance level and state clearly your conclusion.

3 The product moment correlation coefficient between the brain mass and liver mass for a random sample of 15 mice is found. This value is to be tested using a 5% significance level to determine whether there is an association between the masses in mice.

Section 10.1

(a) Within what range of values would it be necessary for r, the sample correlation coefficient, to lie in order to conclude that an association does indeed exist between brain and liver mass?

(b) What is the meaning of:
(i) a Type I error,
(ii) a Type II error,

in the context of this question ?

4 In each of the two cases given below, a correlation coefficient is going to be evaluated in order to determine whether there is an association between the two variables involved. State, for each example, whether the Spearman's rank or the product moment coefficient would be the appropriate measure to evaluate.

Section 10.2

(a) Ranking of two ski jumps for style in a competition with six competitors.

Person	1	2	3	4	5	6
Jump A	4	1	6	2	3	5
Jump B	3	2	6	5	1	4

(b) Length of two ski jumps in a competition with six competitors.

Person	1	2	3	4	5	6
Length A	84	91	86	94	82	79
Length B	91	88	78	91	87	85

10

Test yourself (continued)	What to review

5 A tea taster is given seven infusions of tea leaves to put in order of preference. The price per kilo of the leaves is also put in rank order with the following results:

Sections 10.2 and 10.3

Tea	Preference	Price
A	4	3
B	2.5	1
C	1	2
D	6	6
E	2.5	4
F	7	5
G	5	7

  (a) Why do the ranks 2.5 appear in the preference column for teas B and E?

  (b) Calculate the value of the Spearman's rank correlation coefficient between preference and price.

  (c) Comment on your result.

6 Suggest a reason why a value for the Spearman's rank correlation coefficient for a set of paired data may be much higher than the value of the product moment coefficient for the same data.

Section 10.5

7 Assign rank order values to the following data where it is appropriate to do so. If the data is already suitably ranked, it should be left in this form.

Section 10.1

  (a)

Pupil	1	2	3	4	5	6
Drama grade	A	A+	C	D	B	E
Exam score	73	82	46	55	61	44

  (b)

Cake	1	2	3	4	5	6
Price	49	85	65	75	45	50
Preference rank	3	2	6	5	1	4

8 Find the Spearman's rank correlation coefficient between:

  (a) the drama grade and exam score in question **7(a)**;

  (b) the price and the preference rank in question **7(b)**.

  (c) Stating the null and alternative hypotheses used, carry out a test to determine whether, at the 5% significance level, there is evidence of association between:

    (i) drama grade and exam score,   **(ii)** price and preference.

Section 10.4

1 $H_0 \; \rho = 0$, $\alpha = 0.05$,   1 tail.
    $H_1 \; \rho > 0$, $n = 12$,   ts $= 0.874$,   cv $= 0.497$.
    Reject H_0: significant evidence of direct association between age and maintenance costs.

2 (a) $H_0 \; \rho = 0$, $H_1 \; \rho < 0$;
    (b) ts $= -0.675$, $\alpha = 0.01$,   cv $= -0.592$,   1 tail $n = 15$.
    Reject H_0: significant evidence of inverse association between sociology and research scores.

3 $n = 15$
    (a) $\alpha = 0.05$,   2 tail,   cv $= -0.514$.
    range $-1 \leqslant r \leqslant -0.514,\; 0.514 \leqslant r \leqslant 1$.
    (b) (i) Type 1 error → conclude an association does exist between brain mass and liver mass when, in fact, it does not exist,
         (ii) Type 2 error → conclude no association exists between brain mass and liver mass when, in fact, an association does exist.

4 (a) Spearman's;   **(b)** PMCC.

5 (a) B/E both equal preferences. The rank values 2 and 3 are shared between them;
    (b) 0.739 (or 0.741 using Σd^2);   **(c)** Reasonable agreement between preference and price.

6 Non-linear relationship.
Data closely connected in rank order but not following a straight line relationship.

7 (a)

Pupil	1	2	3	4	5	6
Drama	2	1	4	5	3	6
Exam	2	1	5	4	3	6

(b)

Cake	1	2	3	4	5	6
Price	5	1	3	2	6	4
Pref	3	2	6	5	1	4

8 (a) 0.943;   **(b)** -0.371;
    (c) H_0 pop ranks independent, H_1 pop ranks not independent, $n = 6$, two-tail, 5%;
       (i) ts $= 0.943$,   cv $= \pm 0.8286$,   Reject H_0: significant evidence of association between drama grade and exam score (association is positive)
       (ii) ts $= -0.371$,   cv $= \pm 0.8286$ Accept H_0: insufficient evidence of association between price and preference rank.

01

Exam style practice papers

SS02

Time allowed 1 hour 30 minutes

Answer **all** questions

---

1 The following table shows the expenditure on sea travel, in £million, by UK households. The figures have been adjusted to constant 1995 prices. These should be used throughout the question.

Year	1998	1999
Quarter 1	163	165
Quarter 2	265	293
Quarter 3	409	469
Quarter 4	186	196

(a) Plot the data together with a suitable moving average. (7 marks)

(b) Predict the expenditure in Quarter 1 2000. Indicate the method you have used. (5 marks)

(c) Comment on your method of forecasting, given that the actual figure for Quarter 1, 2000 was £151 million. (3 marks)

2 The number of telephone calls to a university admissions office during working hours may be modelled by a Poisson distribution with mean 1.4 per minute.

(a) Find the probability that the number of calls received during a particular:
  (i) 1-minute interval is 3 or fewer; (2 marks)
  (ii) 1-minute interval is exactly 3; (2 marks)
  (iii) 5-minute interval is more than five. (3 marks)

(b) In a particular 1-minute interval, how many calls will be exceeded with probability just greater than 0.05? (2 marks)

(c) Explain why the Poisson distribution is unlikely to be a suitable model for:

(i) the number of calls received by the admissions office per minute over a 24-hour period; (2 marks)

(ii) the number of calls answered per minute during a busy period when the telephone is often engaged. (2 marks)

3 Applicants for an assembly job take a test of manual dexterity. Current employees took an average of 60 seconds to complete the test. The times, in seconds, taken to complete the test by a random sample of applicants were:

63   125   77   49   74   67   59   66   102

(a) Stating clearly your null and alternative hypotheses, investigate, at the 5% significance level, whether the mean time taken by applicants to complete the test is greater than that taken by current employees. Assume that the times are normally distributed with standard deviation 16 seconds. (8 marks)

(b) It later emerged that due to a misunderstanding of how the test should be timed all the recorded times for the new applicants were five seconds too long. Does this further information affect your conclusion in **(a)**? (4 marks)

4 Travellers (excluding season ticket holders) on a city centre tram must purchase tickets from a machine on the platform which only accepts coins. The cost, £X, of a ticket depends on the destination of the journey and may be modelled by the following probability distribution:

x	P(X = x)
0.8	0.26
1.2	0.34
1.6	0.18
2.0	0.10
2.8	0.12

(a) Calculate:

(i) the mean of X;

(ii) $E(X^2)$;

(iii) the standard deviation of X. (6 marks)

(b) A visitor to the city who is not aware of the need to pay with coins arrives on the platform carrying £1.44 in change. Find the probability the visitor is able to buy an appropriate ticket. Assume that the distribution adequately models the cost of the ticket required by the visitor. (1 mark)

(c) Explain why the answer to **(b)** might not apply to a regular (non-season ticket holder) user of the tram who is carrying £1.44 in change. (2 marks)

5   **Parliamentary elections**

Thousands

	25 Oct 1951	26 May 1955	8 Oct 1959	15 Oct 1964	31 Mar 1966	18 June 1970*	28 Feb 1974	10 Oct 1974	3 May 1979	9 June 1983	11 June 1987	9 April 1992	1 May 1997‡
United Kingdom													
Number of electors	34 919	34 852	35 397	35 894	35 957	39 615	40 256	40 256	41 573	42 704	43 666	43 719	43 830
Average-electors per seat	55.9	55.3	56.2	57.0	57.1	62.9	63.4	63.4	65.5	66.7	67.2	67.2	
Number of valid votes counted	28 597	26 760	27 863	27 657	27 265	28 345	31 340	29 189	31 221	30 671	32 530	33 551	31 286
As percentage of electorate	81.9	76.8	78.7	77.1	75.8	71.5	77.9	72.5	75.1	71.8	74.5	76.7	71.4
Number of Members of Parliament elected:	625	630	630	630	630	630	635	635	635	650	650	651	659
Conservative	320	344	364	303	253	330	296	276	339	396	375	336	
Labour	295	277	258	317	363	287	301	319	268	209	229	271	418
Liberal Democrat	6	6	6	9	12	6	14	13	11	17	17	20	46
Social Democratic Party	–	–	–	–	–	–	–	–	–	6	5	–	–
Scottish National Party	–	–	–	–	–	1	7	11	2	2	3	3	6
Plaid Cymru	–	–	–	–	–	–	2	3	2	2	3	4	4
Other†	4	3	2	1	2	6	15	13	13	18	18	17	20

* The Representation of the People Act 1969 lowered the minimum voting age from 21 to 18 years with effect from 16 February 1970.

† Including the Speaker.

‡ Provisional.

Source: Home Office.

(a) How many valid votes were counted at the election on 9th June 1983?   **(2 marks)**

(b) Some of the figures in the last column have been obliterated.

    (i) How many Conservative Members of Parliament were elected on 1st May 1997?

    (ii) What was the average number of electors per seat at the election on 1st May 1997?   **(5 marks)**

(c) Describe, briefly, the trend of

    (i) 'Number of electors'

    (ii) 'Number of valid votes counted.'

  Compare these two trends.   **(6 marks)**

6 A supermarket employs Agyeman to conduct a survey into customer satisfaction. On a particular day Agyeman chooses a random number, r, between 1 and 12. He then asks the rth customer and every 12th customer thereafter to complete a questionnaire until he has asked a total of 50 customers.

(a) What is the name given to this method of sampling?   **(1 mark)**

(b) Are all the first 600 customers entering the supermarket that day equally likely to be included in the sample? Explain your answer.   **(3 marks)**

(c) Is the chosen sample a random sample of the first 600 customers entering the supermarket? Explain your answer.   **(3 marks)**

(d) The total number of customers entering the supermarket that day was 1912. If Agyeman had asked the rth customer (selected as before), and thereafter every 100th customer entering the supermarket, would all customers entering the supermarket that day be equally likely to be included in this sample? Explain your answer. (3 marks)

(e) Discuss briefly whether the views expressed by the customers selected by either method are likely to be representative of the views of all customers. (3 marks)

SS03

Time allowed 1 hour 30 minutes

Answer **all** questions

1 A large smelting plant is situated by a river. Environmental groups are concerned that waste from this plant may be leading to pollution of the river and a series of readings are taken from various accessible sites along the river.

The distance down river from the plant is recorded together with the measure of pH of the water sample taken from this river. A pH value of about 7 is regarded as neutral which is desirable. Readings below 7 indicate acidity and readings above 7 indicate alkalinity. The values obtained are given in the following table:

X Distance (km)	y (pH)
9.4	6.4
6.5	5.0
5.1	4.9
1.6	4.4
2.5	4.6
4.1	4.7
8.6	5.3
10.2	7.4
3.5	4.7
4.5	4.8

(a) Plot a scatter diagram to illustrate the above data and comment on any noticeable features. (5 marks)

(b) Calculate Spearman's rank correlation coefficient. (7 marks)

(c) Stating clearly your null and alternative hypotheses, investigate whether the pH value is associated with the distance. Use a 5% significance level. Interpret your conclusion in the context of the question. (4 marks)

(d) The value of the product moment correlation coefficient is 0.880. Explain why you might expect the value of the product correlation coefficient for this data to be substantially different from the value obtained for Spearman's rank correlation coefficient in **(b)**. (2 marks)

2 A random sample of adults from three different areas of a city was asked whether they would prefer more money to be spent on public services or a reduction in the tax on petrol. The replies are summarised in the following table:

	Area		
	A	**B**	**C**
More money on public services	44	52	54
Reduce petrol tax	46	28	26

(a) Investigate, at the 5% significance level, whether the answer to the question is independent of the area. (9 marks)

(b) The following table shows the answers, in **percentages**, classified by whether or not the respondent had access to a car for their personal use.

	Access to car	**No access to car**
More money on public services	24	36
Reduce petrol tax	34	6

    (i) Form the data into a contingency table suitable for analysing using the χ^2 distribution. (2 marks)

    (ii) Investigate, at the 5% significance level, whether the answer to the question is independent of whether or not the adult has access to a private car for their personal use. (9 marks)

(c) Interpret your results. (3 marks)

3 A random selection of 12 students is made from first year science students at a university in London. These 12 students all take two modules in the second term of their course. After attending these two modules, the students complete an assessment sheet which enables an evaluation of each module to be made.

The two modules have the same level of difficulty and both involve a similar amount of understanding and numerical work.

Module 1 is taught by traditional lecture methods with support offered in smaller groups of around ten students each on three occasions during the 10-week term.

Module 2 is taught using a series of assignments which students complete on their own with weekly tutorial support in a group of five students all term.

The results of the completed evaluations were as follows:

Student	1	2	3	4	5	6	7	8	9	10	11	12
Module 1	45	29	72	52	55	39	85	62	51	35	50	46
Module 2	67	44	75	60	59	34	78	58	64	41	93	57

(a) Explain the advantages of using paired data in testing whether there is any difference between students' assessments of the two modules. (2 marks)

(b) Carry out a Wilcoxon signed-rank test, using a 5% level of significance, to test whether there is any difference between the students' evaluation of the teaching methods for the two modules. (10 marks)

(c) At the end of their first year, the students all take five exams and the results of these, together with the results from four pieces of coursework, make up a final first year assessment for each student. From past experience, it is expected that the median mark for this assessment at the end of year 1 will be 70%.

The end of year results for the 12 students involved in this trial were:

$$64 \quad 60 \quad 58 \quad 64 \quad 59 \quad 66 \quad 61 \quad 50 \quad 66 \quad 68 \quad 72 \quad 88$$

Carry out a sign test, at the 5% level of significance, to determine whether the first year students have performed below the level expected. (7 marks)

(d) How might your results in (c) affect your conclusion to (b)? (2 marks)

4 The table below shows the percentage porosity of bricks fired at three different temperatures.

1400 °C	1450 °C	1500 °C
18.8	15.7	15.0
14.7	14.3	13.1
14.9	12.5	11.5
14.2	12.6	11.7
15.5	12.2	11.3
15.6		
15.8		

(a) Use an appropriate test to examine, at the 5% significance level, the hypothesis that all three samples come from identical populations. (10 marks)

(b) Interpret your conclusion in the context of the question. (3 marks)

Appendix

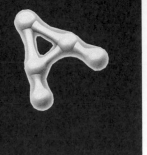

Table 1 Cumulative binomial distribution function

The tabulated value is $P(X \leq x)$, where X has a binomial distribution with parameters n and p.

x \ p	0.01	0.02	0.03	0.04	0.05	0.06	0.07	0.08	0.09	0.10	0.15	0.20	0.25	0.30	0.35	0.40	0.45	0.50	p \ x
n = 2 0	0.9801	0.9604	0.9409	0.9216	0.9025	0.8836	0.8649	0.8464	0.8281	0.8100	0.7225	0.6400	0.5625	0.4900	0.4225	0.3600	0.3025	0.2500	0
1	0.9999	0.9996	0.9991	0.9984	0.9975	0.9964	0.9951	0.9936	0.9919	0.9900	0.9775	0.9600	0.9375	0.9100	0.8775	0.8400	0.7975	0.7500	1
2	1.0000	1.0000	1.0000	1.0000	1.0000	1.0000	1.0000	1.0000	1.0000	1.0000	1.0000	1.0000	1.0000	1.0000	1.0000	1.0000	1.0000	1.0000	2
n = 3 0	0.9703	0.9412	0.9127	0.8847	0.8574	0.8306	0.8044	0.7787	0.7536	0.7290	0.6141	0.5120	0.4219	0.3430	0.2746	0.2160	0.1664	0.1250	0
1	0.9997	0.9988	0.9974	0.9953	0.9928	0.9896	0.9860	0.9818	0.9772	0.9720	0.9393	0.8960	0.8438	0.7840	0.7183	0.6480	0.5747	0.5000	1
2	1.0000	1.0000	1.0000	0.9999	0.9999	0.9998	0.9997	0.9995	0.9993	0.9990	0.9966	0.9920	0.9844	0.9730	0.9571	0.9360	0.9089	0.8750	2
3				1.0000	1.0000	1.0000	1.0000	1.0000	1.0000	1.0000	1.0000	1.0000	1.0000	1.0000	1.0000	1.0000	1.0000	1.0000	3
n = 4 0	0.9606	0.9224	0.8853	0.8493	0.8145	0.7807	0.7481	0.7164	0.6857	0.6561	0.5220	0.4096	0.3164	0.2401	0.1785	0.1296	0.0915	0.0625	0
1	0.9994	0.9977	0.9948	0.9909	0.9860	0.9801	0.9733	0.9656	0.9570	0.9477	0.8905	0.8192	0.7383	0.6517	0.5630	0.4752	0.3910	0.3125	1
2	1.0000	1.0000	0.9999	0.9998	0.9995	0.9992	0.9987	0.9981	0.9973	0.9963	0.9880	0.9728	0.9492	0.9163	0.8735	0.8208	0.7585	0.6875	2
3			1.0000	1.0000	1.0000	1.0000	1.0000	1.0000	0.9999	0.9999	0.99995	0.9984	0.9961	0.9919	0.9850	0.9744	0.9590	0.9375	3
4									1.0000	1.0000	1.0000	1.0000	1.0000	1.0000	1.0000	1.0000	1.0000	1.0000	4
n = 5 0	0.9510	0.9039	0.8587	0.8154	0.7738	0.7339	0.6957	0.6591	0.6240	0.5905	0.4437	0.3277	0.2373	0.1681	0.1160	0.0778	0.0503	0.0313	0
1	0.9990	0.9962	0.9915	0.9852	0.9774	0.9681	0.9575	0.9456	0.9326	0.9185	0.8352	0.7373	0.6328	0.5282	0.4284	0.3370	0.2562	0.1875	1
2	1.0000	0.9999	0.9997	0.9994	0.9988	0.9980	0.9969	0.9955	0.9937	0.9914	0.9734	0.9421	0.8965	0.8369	0.7648	0.6826	0.5931	0.5000	2
3			1.0000	1.0000	1.0000	0.9999	0.9999	0.9998	0.9997	0.9995	0.9978	0.9933	0.9844	0.9692	0.9460	0.9130	0.8688	0.8125	3
4							1.0000	1.0000	1.0000	1.0000	0.9999	0.9997	0.9990	0.9976	0.9947	0.9898	0.9815	0.9688	4
5											1.0000	1.0000	1.0000	1.0000	1.0000	1.0000	1.0000	1.0000	5
n = 6 0	0.9415	0.8858	0.8330	0.7828	0.7351	0.6899	0.6470	0.6064	0.5679	0.5314	0.3771	0.2621	0.1780	0.1176	0.0754	0.0467	0.0277	0.0156	0
1	0.9985	0.9943	0.9875	0.9784	0.9672	0.9541	0.9392	0.9227	0.9048	0.8857	0.7765	0.6554	0.5339	0.4202	0.3191	0.2333	0.1636	0.1094	1
2	1.0000	0.9998	0.9995	0.9988	0.9978	0.9962	0.9942	0.9915	0.9882	0.9842	0.9527	0.9011	0.8306	0.7443	0.6471	0.5443	0.4415	0.3438	2
3		1.0000	1.0000	1.0000	0.9999	0.9998	0.9997	0.9995	0.9992	0.9987	0.9941	0.9830	0.9624	0.9295	0.8826	0.8208	0.7447	0.6563	3
4						1.0000	1.0000	1.0000	1.0000	0.9999	0.9996	0.9984	0.9954	0.9891	0.9777	0.9590	0.9308	0.8906	4
5										1.0000	1.0000	0.9999	0.9998	0.9993	0.9982	0.9959	0.9917	0.9844	5
6											1.0000	1.0000	1.0000	1.0000	1.0000	1.0000	1.0000	1.0000	6
n = 7 0	0.9321	0.8681	0.8080	0.7514	0.6983	0.6485	0.6017	0.5578	0.5168	0.4783	0.3206	0.2097	0.1335	0.0824	0.0490	0.0280	0.0152	0.0078	0
1	0.9980	0.9921	0.9829	0.9706	0.9556	0.9382	0.9187	0.8974	0.8745	0.8503	0.7166	0.5767	0.4449	0.3294	0.2338	0.1586	0.1024	0.0625	1
2	1.0000	0.9997	0.9991	0.9980	0.9962	0.9937	0.9903	0.9860	0.9807	0.9743	0.9262	0.8520	0.7564	0.6471	0.5323	0.4199	0.3164	0.2266	2
3		1.0000	1.0000	0.9999	0.9998	0.9996	0.9993	0.9988	0.9982	0.9973	0.9879	0.9667	0.9294	0.8740	0.8002	0.7102	0.6083	0.5000	3
4				1.0000	1.0000	1.0000	1.0000	0.9999	0.9999	0.9998	0.9988	0.9953	0.9871	0.9712	0.9444	0.9037	0.8471	0.7734	4
5							1.0000	1.0000	1.0000	1.0000	0.9999	0.9996	0.9987	0.9962	0.9910	0.9812	0.9643	0.9375	5
6												1.0000	0.9999	0.9998	0.9994	0.9984	0.9963	0.9922	6
7													1.0000	1.0000	1.0000	1.0000	1.0000	1.0000	7
n = 8 0	0.9227	0.8508	0.7837	0.7214	0.6634	0.6096	0.5596	0.5132	0.4703	0.4305	0.2725	0.1678	0.1001	0.0576	0.0319	0.0168	0.0084	0.0039	0
1	0.9972	0.9897	0.9777	0.9619	0.9428	0.9208	0.8965	0.8702	0.8423	0.8131	0.6572	0.5033	0.3671	0.2553	0.1691	0.1064	0.0632	0.0352	1
2	0.9999	0.9996	0.9987	0.9969	0.9942	0.9904	0.9853	0.9789	0.9711	0.9619	0.8948	0.7969	0.6785	0.5518	0.4278	0.3154	0.2201	0.1445	2
3	1.0000	1.0000	0.9999	0.9998	0.9996	0.9993	0.9987	0.9978	0.9966	0.9950	0.9786	0.9437	0.8862	0.8059	0.7064	0.5941	0.4770	0.3633	3
4			1.0000	1.0000	1.0000	1.0000	0.9999	0.9999	0.9997	0.9996	0.9971	0.9896	0.9727	0.9420	0.8939	0.8263	0.7396	0.6367	4
5						1.0000	1.0000	1.0000	1.0000	1.0000	0.9998	0.9988	0.9958	0.9887	0.9747	0.9502	0.9115	0.8555	5
6											1.0000	0.9999	0.9996	0.9987	0.9964	0.9915	0.9819	0.9648	6
7													1.0000	0.9999	0.9998	0.9993	0.9983	0.9961	7
8																1.0000	1.0000	1.0000	8

Table 1 Cumulative binomial distribution function (cont.)

x \ p	0.01	0.02	0.03	0.04	0.05	0.06	0.07	0.08	0.09	0.10	0.15	0.20	0.25	0.30	0.35	0.40	0.45	0.50	p \ x
n = 9 0	0.9135	0.8337	0.7602	0.6925	0.6302	0.5730	0.5204	0.4722	0.4279	0.3874	0.2316	0.1342	0.0751	0.0404	0.0207	0.0101	0.0046	0.0020	0
1	0.9966	0.9869	0.9718	0.9522	0.9288	0.9022	0.8729	0.8417	0.8088	0.7748	0.5995	0.4362	0.3003	0.1960	0.1211	0.0705	0.0385	0.0195	1
2	0.9999	0.9994	0.9980	0.9955	0.9916	0.9862	0.9791	0.9702	0.9595	0.9470	0.8591	0.7382	0.6007	0.4628	0.3373	0.2318	0.1495	0.0898	2
3	1.0000	1.0000	0.9999	0.9997	0.9994	0.9987	0.9977	0.9963	0.9943	0.9917	0.9661	0.9144	0.8343	0.7297	0.6089	0.4826	0.3614	0.2539	3
4			1.0000	1.0000	1.0000	0.9999	0.9998	0.9997	0.9995	0.9991	0.9944	0.9804	0.9511	0.9012	0.8283	0.7334	0.6214	0.5000	4
5						1.0000	1.0000	1.0000	1.0000	0.9999	0.9994	0.9969	0.9900	0.9747	0.9464	0.9006	0.8342	0.7461	5
6										1.0000	1.0000	0.9997	0.9987	0.9957	0.9888	0.9750	0.9502	0.9102	6
7												1.0000	0.9999	0.9996	0.9986	0.9962	0.9909	0.9805	7
8													1.0000	1.0000	0.9999	0.9997	0.9992	0.9980	8
9															1.0000	1.0000	1.0000	1.0000	9
n = 10 0	0.9044	0.8171	0.7374	0.6648	0.5987	0.5386	0.4840	0.4344	0.3894	0.3487	0.1969	0.1074	0.0563	0.0282	0.0135	0.0060	0.0025	0.0010	0
1	0.9957	0.9838	0.9655	0.9418	0.9139	0.8824	0.8483	0.8121	0.7746	0.7361	0.5443	0.3758	0.2440	0.1493	0.0860	0.0464	0.0233	0.0107	1
2	0.9999	0.9991	0.9972	0.9938	0.9885	0.9812	0.9717	0.9599	0.9460	0.9298	0.8202	0.6778	0.5256	0.3828	0.2616	0.1673	0.0996	0.0547	2
3	1.0000	1.0000	0.9999	0.9996	0.9990	0.9980	0.9964	0.9942	0.9912	0.9872	0.9500	0.8791	0.7759	0.6496	0.5138	0.3823	0.2660	0.1719	3
4			1.0000	1.0000	0.9999	0.9998	0.9997	0.9994	0.9990	0.9984	0.9901	0.9672	0.9219	0.8497	0.7515	0.6331	0.5044	0.3770	4
5					1.0000	1.0000	1.0000	1.0000	0.9999	0.9999	0.9986	0.9936	0.9803	0.9527	0.9051	0.8338	0.7384	0.6230	5
6										1.0000	0.9999	0.9991	0.9965	0.9894	0.9740	0.9452	0.8980	0.8281	6
7											1.0000	0.9999	0.9996	0.9984	0.9952	0.9877	0.9726	0.9453	7
8												1.0000	1.0000	0.9999	0.9995	0.9983	0.9955	0.9893	8
9														1.0000	1.0000	0.9999	0.9997	0.9990	9
10																1.0000	1.0000	1.0000	10
n = 11 0	0.8953	0.8007	0.7153	0.6382	0.5688	0.5063	0.4501	0.3996	0.3544	0.3138	0.1673	0.0859	0.0422	0.0198	0.0088	0.0036	0.0014	0.0005	0
1	0.9948	0.9805	0.9587	0.9308	0.8981	0.8618	0.8228	0.7819	0.7399	0.6974	0.4922	0.3221	0.1971	0.1130	0.0606	0.0302	0.0139	0.0059	1
2	0.9998	0.9988	0.9963	0.9917	0.9848	0.9752	0.9630	0.9481	0.9305	0.9104	0.7788	0.6174	0.4552	0.3127	0.2001	0.1189	0.0652	0.0327	2
3	1.0000	1.0000	0.9998	0.9993	0.9984	0.9970	0.9947	0.9915	0.9871	0.9815	0.9306	0.8389	0.7133	0.5696	0.4256	0.2963	0.1911	0.1133	3
4			1.0000	1.0000	0.9999	0.9997	0.9995	0.9990	0.9983	0.9972	0.9841	0.9496	0.8854	0.7897	0.6683	0.5328	0.3971	0.2744	4
5					1.0000	1.0000	1.0000	0.9999	0.9998	0.9997	0.9973	0.9883	0.9657	0.9218	0.8513	0.7535	0.6331	0.5000	5
6							1.0000	1.0000	1.0000	0.9999	0.9997	0.9980	0.9924	0.9784	0.9499	0.9006	0.8262	0.7256	6
7											1.0000	0.9998	0.9988	0.9957	0.9878	0.9707	0.9390	0.8667	7
8												1.0000	0.9999	0.9994	0.9980	0.9941	0.9852	0.9673	8
9													1.0000	1.0000	0.9998	0.9993	0.9978	0.9941	9
10															1.0000	1.0000	0.9998	0.9995	10
11																		1.0000	11
n = 12 0	0.8864	0.7847	0.6938	0.6127	0.5404	0.4759	0.4186	0.3677	0.3225	0.2824	0.1422	0.0687	0.0317	0.0138	0.0057	0.0022	0.0008	0.0002	0
1	0.9938	0.9769	0.9514	0.9191	0.8816	0.8405	0.7967	0.7513	0.7052	0.6590	0.4435	0.2749	0.1584	0.0850	0.0424	0.0196	0.0083	0.0032	1
2	0.9998	0.9985	0.9952	0.9893	0.9804	0.9684	0.9532	0.9348	0.9134	0.8891	0.7358	0.5583	0.3907	0.2528	0.1513	0.0834	0.0421	0.0193	2
3	1.0000	0.9999	0.9997	0.9990	0.9978	0.9957	0.9925	0.9880	0.9820	0.9744	0.9078	0.7946	0.6488	0.4925	0.3467	0.2253	0.1345	0.0730	3
4	1.0000	1.0000	1.0000	0.9999	0.9998	0.9996	0.9991	0.9984	0.9973	0.9957	0.9761	0.9274	0.8424	0.7237	0.5833	0.4382	0.3044	0.1938	4
5				1.0000	1.0000	1.0000	0.9999	0.9998	0.9997	0.9995	0.9954	0.9806	0.9456	0.8822	0.7873	0.6652	0.5269	0.3872	5
6							1.0000	1.0000	1.0000	0.9999	0.9993	0.9961	0.9857	0.9614	0.9154	0.8418	0.7393	0.6128	6
7										1.0000	0.9999	0.9994	0.9972	0.9905	0.9745	0.9427	0.8883	0.8062	7
8											1.0000	0.9999	0.9996	0.9983	0.9944	0.9847	0.9644	0.9270	8
9												1.0000	1.0000	0.9998	0.9992	0.9972	0.9921	0.9807	9
10														1.0000	0.9999	0.9997	0.9989	0.9968	10
11															1.0000	1.0000	0.9999	0.9998	11
12																	1.0000	1.0000	12
n = 13 0	0.8775	0.7690	0.6730	0.5882	0.5133	0.4474	0.3893	0.3383	0.2935	0.2542	0.1209	0.0550	0.0238	0.0097	0.0037	0.0013	0.0004	0.0001	0
1	0.9928	0.9730	0.9436	0.9068	0.8646	0.8186	0.7702	0.7206	0.6707	0.6213	0.3983	0.2336	0.1267	0.0637	0.0296	0.0126	0.0049	0.0017	1
2	0.9997	0.9980	0.9938	0.9865	0.9755	0.9608	0.9422	0.9201	0.8946	0.8661	0.6920	0.5017	0.3326	0.2025	0.1132	0.0579	0.0269	0.0112	2
3	1.0000	0.9999	0.9995	0.9986	0.9969	0.9940	0.9897	0.9837	0.9758	0.9658	0.8820	0.7473	0.5843	0.4206	0.2783	0.1686	0.0929	0.0461	3
4		1.0000	1.0000	0.9999	0.9997	0.9993	0.9987	0.9976	0.9959	0.9935	0.9658	0.9009	0.7940	0.6543	0.5005	0.3530	0.2279	0.1334	4
5						1.0000	1.0000	0.9999	0.9997	0.9991	0.9925	0.9700	0.9198	0.8346	0.7159	0.5744	0.4268	0.2905	5
6									1.0000	0.9999	0.9987	0.9930	0.9757	0.9376	0.8705	0.7712	0.6437	0.5000	6
7									1.0000	1.0000	0.9998	0.9988	0.9944	0.9818	0.9538	0.9023	0.8212	0.7095	7
8											1.0000	0.9998	0.9990	0.9960	0.9874	0.9679	0.9302	0.8666	8
9												1.0000	0.9999	0.9993	0.9975	0.9922	0.9797	0.9539	9
10													1.0000	0.9999	0.9997	0.9987	0.9959	0.9888	10
11														1.0000	1.0000	0.9999	0.9995	0.9983	11
12																1.0000	1.0000	0.9999	12
13																		1.0000	13

Table 1 Cumulative binomial distribution function (cont.)

n = 14

x	0.01	0.02	0.03	0.04	0.05	0.06	0.07	0.08	0.09	0.10	0.15	0.20	0.25	0.30	0.35	0.40	0.45	0.50	x
0	0.8687	0.7536	0.6528	0.5647	0.4877	0.4205	0.3620	0.3112	0.2670	0.2288	0.1028	0.0440	0.0178	0.0068	0.0024	0.0008	0.0002	0.0001	0
1	0.9926	0.9690	0.9355	0.8941	0.8470	0.7963	0.7436	0.6900	0.6368	0.5846	0.3567	0.1979	0.1010	0.0475	0.0205	0.0081	0.0029	0.0009	1
2	0.9997	0.9975	0.9923	0.9833	0.9699	0.9522	0.9302	0.9042	0.8745	0.8416	0.6479	0.4481	0.2811	0.1608	0.0839	0.0398	0.0170	0.0065	2
3	1.0000	0.9999	0.9994	0.9981	0.9958	0.9920	0.9864	0.9786	0.9685	0.9559	0.8535	0.6982	0.5213	0.3552	0.2205	0.1243	0.0632	0.0287	3
4		1.0000	1.0000	0.9998	0.9996	0.9990	0.9980	0.9965	0.9941	0.9908	0.9533	0.8702	0.7415	0.5842	0.4227	0.2793	0.1672	0.0898	4
5				1.0000	1.0000	0.9999	0.9998	0.9996	0.9992	0.9985	0.9885	0.9561	0.8883	0.7805	0.6405	0.4859	0.3373	0.2120	5
6						1.0000	1.0000	1.0000	0.9999	0.9998	0.9978	0.9884	0.9617	0.9067	0.8164	0.6925	0.5461	0.3953	6
7								1.0000	1.0000	1.0000	0.9997	0.9976	0.9897	0.9685	0.92427	0.8499	0.7414	0.6047	7
8											1.0000	0.9996	0.9978	0.9917	0.9757	0.9417	0.8811	0.7880	8
9												1.0000	0.9997	0.9983	0.9940	0.9825	0.9574	0.9102	9
10													1.0000	0.9998	0.9989	0.9961	0.9886	0.9713	10
11														1.0000	0.9999	0.9994	0.9978	0.9935	11
12															1.0000	0.9999	0.9997	0.9991	12
13																1.0000	1.0000	0.9999	13
14																		1.0000	14

n = 15

x	0.01	0.02	0.03	0.04	0.05	0.06	0.07	0.08	0.09	0.10	0.15	0.20	0.25	0.30	0.35	0.40	0.45	0.50	x
0	0.8601	0.7386	0.6333	0.5421	0.4633	0.3953	0.3367	0.2863	0.2430	0.2059	0.0874	0.0352	0.0134	0.0047	0.0016	0.0005	0.0001	0.0000	0
1	0.9904	0.9647	0.9270	0.8809	0.8290	0.7738	0.7168	0.6597	0.6035	0.5490	0.3186	0.1671	0.0802	0.0353	0.0142	0.0052	0.0017	0.0005	1
2	0.9996	0.9970	0.9906	0.9797	0.9638	0.9429	0.9171	0.8870	0.8531	0.8159	0.6042	0.3980	0.2361	0.1268	0.0617	0.0271	0.0107	0.0037	2
3	1.0000	0.9998	0.9992	0.9976	0.9945	0.9896	0.9825	0.9727	0.9601	0.9444	0.8227	0.6482	0.4613	0.2969	0.1727	0.0905	0.0424	0.0176	3
4		1.0000	0.9999	0.9998	0.9994	0.9986	0.9972	0.9950	0.9918	0.9873	0.9383	0.8358	0.6865	0.5155	0.3519	0.2173	0.1204	0.0592	4
5			1.0000	1.0000	0.9999	0.9999	0.9997	0.9993	0.9987	0.9978	0.9832	0.9389	0.8516	0.7216	0.5643	0.4032	0.2608	0.1509	5
6					1.0000	1.0000	1.0000	0.9999	0.9998	0.9997	0.9964	0.9819	0.9434	0.8689	0.7548	0.6098	0.4522	0.3036	6
7								1.0000	1.0000	1.0000	0.9994	0.9958	0.9827	0.9500	0.8868	0.7869	0.6535	0.5000	7
8											0.9999	0.9992	0.9958	0.9848	0.9578	0.9050	0.8182	0.6964	8
9											1.0000	0.9999	0.9992	0.9963	0.9876	0.9662	0.9231	0.8491	9
10												1.0000	0.9999	0.9993	0.9972	0.9907	0.9745	0.9408	10
11													1.0000	0.9999	0.9995	0.9981	0.9937	0.9824	11
12														1.0000	0.9999	0.9997	0.9989	0.9963	12
13															1.0000	1.0000	0.9999	0.9995	13
14																	1.0000	1.0000	14

n = 20

x	0.01	0.02	0.03	0.04	0.05	0.06	0.07	0.08	0.09	0.10	0.15	0.20	0.25	0.30	0.35	0.40	0.45	0.50	x
0	0.8179	0.6676	0.5438	0.4420	0.3585	0.2901	0.2342	0.1887	0.1516	0.1216	0.0388	0.0115	0.0032	0.0008	0.0002	0.0000	0.0000	0.0000	0
1	0.9831	0.9401	0.8802	0.8103	0.7358	0.6605	0.5869	0.5169	0.4516	0.3917	0.1756	0.0692	0.0243	0.0076	0.0021	0.0005	0.0001	0.0000	1
2	0.9990	0.9929	0.9790	0.9561	0.9245	0.8850	0.8390	0.7879	0.7334	0.6769	0.4049	0.2061	0.0913	0.0355	0.0121	0.0036	0.0009	0.0002	2
3	1.0000	0.9994	0.9973	0.9926	0.9841	0.9710	0.9529	0.9294	0.9007	0.8670	0.6477	0.4114	0.2252	0.1071	0.0444	0.0160	0.0049	0.0013	3
4		1.0000	0.9997	0.9990	0.9974	0.9944	0.9893	0.9817	0.9710	0.9568	0.8298	0.6296	0.4148	0.2375	0.1182	0.0510	0.0189	0.0059	4
5			1.0000	0.9999	0.9997	0.9991	0.9981	0.9962	0.9932	0.9887	0.9327	0.8042	0.6172	0.4164	0.2454	0.1256	0.0553	0.0207	5
6				1.0000	1.0000	0.9999	0.9997	0.9994	0.9987	0.9976	0.9781	0.9133	0.7858	0.6080	0.4166	0.2500	0.1299	0.0577	6
7						1.0000	1.0000	0.9999	0.9998	0.9996	0.9941	0.9679	0.8982	0.7723	0.6010	0.4159	0.2520	0.1316	7
8								1.0000	1.0000	0.9999	0.9987	0.9900	0.9591	0.8867	0.7624	0.5956	0.4143	0.2517	8
9										1.0000	0.9998	0.9974	0.9861	0.9520	0.8782	0.7553	0.5914	0.4119	9
10											1.0000	0.9994	0.9961	0.9829	0.9468	0.8725	0.7507	0.5881	10
11												0.9999	0.9991	0.9949	0.9804	0.9435	0.8692	0.7483	11
12												1.0000	0.9998	0.9987	0.9940	0.9790	0.9420	0.8684	12
13													1.0000	0.9997	0.9985	0.9935	0.9786	0.9423	13
14														1.0000	0.9997	0.9984	0.9936	0.9793	14
15															1.0000	0.9997	0.9985	0.9941	15
16																1.0000	0.9997	0.9987	16
17																	1.0000	0.9998	17
18																		1.0000	18

n = 25

x	0.01	0.02	0.03	0.04	0.05	0.06	0.07	0.08	0.09	0.10	0.15	0.20	0.25	0.30	0.35	0.40	0.45	0.50	x
0	0.7778	0.6035	0.4670	0.3604	0.2774	0.2129	0.1630	0.1244	0.0946	0.0718	0.0172	0.0038	0.0008	0.0001	0.0000	0.0000	0.0000	0.0000	0
1	0.9742	0.9114	0.8280	0.7358	0.6424	0.5527	0.4696	0.3947	0.3286	0.2712	0.0931	0.0274	0.0070	0.0016	0.0003	0.0001	0.0000	0.0000	1
2	0.9980	0.9868	0.9620	0.9235	0.8729	0.8129	0.7466	0.6768	0.6063	0.5371	0.2537	0.0982	0.0321	0.0090	0.0021	0.0004	0.0001	0.0000	2
3	0.9999	0.9986	0.9938	0.9835	0.9659	0.9402	0.9064	0.8649	0.8169	0.7636	0.4711	0.2340	0.0962	0.0332	0.0097	0.0024	0.0005	0.0001	3
4	1.0000	0.9999	0.9992	0.9972	0.9928	0.9850	0.9726	0.9549	0.9314	0.9020	0.6821	0.4207	0.2137	0.0905	0.0320	0.0095	0.0023	0.0005	4
5		1.0000	0.9999	0.9996	0.9988	0.9969	0.9935	0.9877	0.9790	0.9666	0.8385	0.6167	0.3783	0.1935	0.0826	0.0294	0.0086	0.0020	5
6			1.0000	1.0000	0.9998	0.9995	0.9987	0.9972	0.9946	0.9905	0.9305	0.7800	0.5611	0.3407	0.1734	0.0736	0.0258	0.0073	6
7					1.0000	0.9999	0.9998	0.9995	0.9989	0.9977	0.9745	0.8909	0.7265	0.5118	0.3061	0.1536	0.0639	0.0216	7
8							1.0000	0.9999	0.9998	0.9995	0.9920	0.9532	0.8506	0.6769	0.4668	0.2735	0.1340	0.0539	8
9								1.0000	0.9999	0.9999	0.9979	0.9827	0.9287	0.8106	0.6303	0.4246	0.2424	0.1148	9
10										1.0000	0.9995	0.9944	0.9703	0.9022	0.7712	0.5858	0.3483	0.2122	10
11											0.9999	0.9985	0.9893	0.9558	0.8746	0.7323	0.5426	0.3450	11
12											1.0000	0.9996	0.9966	0.9825	0.9396	0.8462	0.6937	0.5000	12
13												0.9999	0.9991	0.9940	0.9745	0.9222	0.8173	0.6550	13
14												1.0000	0.9998	0.9982	0.9907	0.9656	0.9040	0.7878	14
15													1.0000	0.9995	0.9971	0.9868	0.9560	0.8852	15
16														0.9999	0.9992	0.9957	0.9826	0.9461	16
17														1.0000	0.9998	0.9988	0.9942	0.9784	17
18															1.0000	0.9997	0.9984	0.9927	18
19																0.9999	0.9996	0.9980	19
20																1.0000	0.9999	0.9995	20
21																	1.0000	0.9999	21
22																		1.0000	22

Table 1 Cumulative binomial distribution function (cont.)

x	0.01	0.02	0.03	0.04	0.05	0.06	0.07	0.08	0.09	0.10	0.15	0.20	0.25	0.30	0.35	0.40	0.45	0.50	x
n = 30																			
0	0.7397	0.5455	0.4010	0.2939	0.2146	0.1563	0.1134	0.0820	0.0591	0.0424	0.0076	0.0012	0.0002	0.0000	0.0000	0.0000	0.0000	0.0000	0
1	0.9639	0.8795	0.7731	0.6612	0.5535	0.4555	0.3694	0.2958	0.2343	0.1837	0.0480	0.0105	0.0020	0.0003	0.0000	0.0000	0.0000	0.0000	1
2	0.9967	0.9783	0.9399	0.8831	0.8122	0.7324	0.6487	0.5654	0.4855	0.4114	0.1514	0.0442	0.0106	0.0021	0.0003	0.0000	0.0000	0.0000	2
3	0.9998	0.9971	0.9881	0.9694	0.9392	0.8974	0.8450	0.7842	0.7175	0.6474	0.3217	0.1227	0.0374	0.0093	0.0019	0.0003	0.0000	0.0000	3
4	1.0000	0.9997	0.9982	0.9937	0.9844	0.9685	0.9447	0.9126	0.8723	0.8245	0.5245	0.2552	0.0979	0.0302	0.0075	0.0015	0.0002	0.0000	4
5		1.0000	0.9998	0.9989	0.9967	0.9921	0.9838	0.9707	0.9519	0.9268	0.7106	0.4275	0.2026	0.0766	0.0233	0.0057	0.0011	0.0002	5
6			1.0000	0.9999	0.9994	0.9983	0.9960	0.9918	0.9848	0.9742	0.8474	0.6070	0.3481	0.1595	0.0586	0.0172	0.0040	0.0007	6
7				1.0000	0.9999	0.9997	0.9992	0.9980	0.9959	0.9922	0.9302	0.7608	0.5143	0.2814	0.1238	0.0435	0.0121	0.0026	7
8					1.0000	1.0000	0.9999	0.9996	0.9990	0.9980	0.9722	0.8713	0.6736	0.4315	0.2247	0.0940	0.0312	0.0081	8
9							1.0000	0.9999	0.9998	0.9995	0.9903	0.9389	0.8034	0.5888	0.3575	0.1763	0.0694	0.0214	9
10								1.0000	0.9999	0.9999	0.9971	0.9744	0.8943	0.7304	0.5078	0.2915	0.1350	0.0494	10
11									1.0000	1.0000	0.9992	0.9905	0.9493	0.8407	0.6548	0.4311	0.2327	0.1002	11
12											0.9998	0.9969	0.9784	0.9155	0.7802	0.5785	0.3592	0.1808	12
13											1.0000	0.9991	0.9918	0.9599	0.8737	0.7145	0.5025	0.2923	13
14												0.9998	0.9973	0.9831	0.9348	0.8246	0.6448	0.4278	14
15												0.9999	0.9992	0.9936	0.9699	0.9029	0.7691	0.5722	15
16												1.0000	0.9998	0.9979	0.9876	0.9519	0.8644	0.7077	16
17													0.9999	0.9994	0.9955	0.9788	0.9286	0.8192	17
18													1.0000	0.9998	0.9986	0.9917	0.9666	0.8998	18
19														1.0000	0.9996	0.9971	0.9862	0.9506	19
20															0.9999	0.9991	0.9950	0.9786	20
21															1.0000	0.9998	0.9984	0.9919	21
22																1.0000	0.9996	0.9974	22
23																	0.9999	0.9993	23
24																	1.0000	0.9998	24
25																		1.0000	25
n = 40																			
0	0.6690	0.4457	0.2957	0.1954	0.1285	0.0842	0.0549	0.0356	0.0230	0.0148	0.0015	0.0001	0.0000	0.0000	0.0000	0.0000	0.0000	0.0000	0
1	0.9393	0.8095	0.6615	0.5210	0.3991	0.2990	0.2201	0.1594	0.1140	0.0805	0.0212	0.0015	0.0001	0.0000	0.0000	0.0000	0.0000	0.0000	1
2	0.9925	0.9543	0.8822	0.7855	0.6767	0.5665	0.4625	0.3694	0.2984	0.2228	0.0486	0.0079	0.0010	0.0001	0.0000	0.0000	0.0000	0.0000	2
3	0.9993	0.9918	0.9686	0.9252	0.8619	0.7827	0.6937	0.6007	0.5092	0.4231	0.1302	0.0285	0.0047	0.0006	0.0001	0.0000	0.0000	0.0000	3
4	1.0000	0.9988	0.9933	0.9790	0.9520	0.9104	0.8546	0.7868	0.7103	0.6290	0.2633	0.0759	0.0160	0.0026	0.0003	0.0000	0.0000	0.0000	4
5		0.9999	0.9988	0.9951	0.9861	0.9691	0.9419	0.9033	0.8535	0.7937	0.4325	0.1613	0.0433	0.0086	0.0013	0.0001	0.0000	0.0000	5
6		1.0000	0.9998	0.9990	0.9966	0.9909	0.9801	0.9624	0.9361	0.9005	0.6077	0.2859	0.0962	0.0238	0.0044	0.0006	0.0001	0.0000	6
7			1.0000	0.9998	0.9993	0.9977	0.9942	0.9873	0.9758	0.9581	0.7559	0.4371	0.1820	0.0553	0.0124	0.0021	0.0002	0.0000	7
8				1.0000	0.9999	0.9995	0.9985	0.9963	0.9919	0.9845	0.8646	0.5931	0.2998	0.1110	0.0303	0.0061	0.0009	0.0001	8
9					1.0000	0.9999	0.9997	0.9990	0.9976	0.9949	0.9328	0.7318	0.4395	0.1959	0.0644	0.0156	0.0027	0.0003	9
10						1.0000	0.9999	0.9998	0.9994	0.9985	0.9701	0.8392	0.5839	0.3087	0.1215	0.0352	0.0074	0.0011	10
11							1.0000	0.9999	0.9998	0.9996	0.9880	0.9125	0.7151	0.4406	0.2053	0.0709	0.0179	0.0032	11
12								1.0000	1.0000	0.9999	0.9957	0.9568	0.8209	0.5772	0.3143	0.1285	0.0386	0.0083	12
13										1.0000	0.9986	0.9806	0.8968	0.7032	0.4408	0.2112	0.0751	0.0192	13
14											0.9996	0.9921	0.9456	0.8074	0.5721	0.3174	0.1326	0.0403	14
15											0.9999	0.9971	0.9738	0.8849	0.6946	0.4402	0.2142	0.0769	15
16											1.0000	0.9990	0.9884	0.9367	0.7978	0.5681	0.3185	0.1341	16
17												0.9997	0.9953	0.9680	0.8761	0.6885	0.4391	0.2148	17
18												0.9999	0.9983	0.9852	0.9301	0.7911	0.5651	0.3179	18
19												1.0000	0.9994	0.9937	0.9637	0.8702	0.6844	0.4373	19
20													0.9998	0.9976	0.9827	0.9256	0.7870	0.5627	20
21													1.0000	0.9991	0.9925	0.9608	0.8669	0.6821	21
22														0.9997	0.9970	0.9811	0.9233	0.7852	22
23														0.9999	0.9989	0.9917	0.9595	0.8659	23
24														1.0000	0.9996	0.9966	0.9804	0.9231	24
25															0.9999	0.9988	0.9914	0.9597	25
26															1.0000	0.9996	0.9966	0.9808	26
27																0.9999	0.9988	0.9917	27
28																1.0000	0.9996	0.9968	28
29																	0.9999	0.9989	29
30																	1.0000	0.9997	30
31																		0.9999	31
32																		1.0000	32

Table 1 Cumulative binomial distribution function (cont.)

x	0.01	0.02	0.03	0.04	0.05	0.06	0.07	0.08	0.09	0.10	0.15	0.20	0.25	0.30	0.35	0.40	0.45	0.50	x
n = 50 0	0.6050	0.3642	0.2181	0.1299	0.0769	0.0453	0.0266	0.0155	0.0090	0.0052	0.0003	0.0000	0.0000	0.0000	0.0000	0.0000	0.0000	0.0000	0
1	0.9106	0.7358	0.5553	0.4005	0.2794	0.1900	0.1265	0.0827	0.0532	0.0338	0.0029	0.0002	0.0000	0.0000	0.0000	0.0000	0.0000	0.0000	1
2	0.9862	0.9216	0.8108	0.6767	0.5405	0.4162	0.3108	0.2260	0.1605	0.1117	0.0142	0.0013	0.0001	0.0000	0.0000	0.0000	0.0000	0.0000	2
3	0.9984	0.9822	0.9372	0.8609	0.7604	0.6473	0.5327	0.4253	0.3303	0.2503	0.0460	0.0057	0.0005	0.0000	0.0000	0.0000	0.0000	0.0000	3
4	0.9999	0.9968	0.9832	0.9510	0.8964	0.8206	0.7290	0.6290	0.5277	0.4312	0.1121	0.0185	0.0021	0.0002	0.0000	0.0000	0.0000	0.0000	4
5	1.0000	0.9995	0.9963	0.9856	0.9622	0.9224	0.8650	0.7919	0.7072	0.6161	0.2194	0.0480	0.0070	0.0007	0.0001	0.0000	0.0000	0.0000	5
6		0.9999	0.9993	0.9964	0.9882	0.9711	0.9417	0.8981	0.8404	0.7702	0.3613	0.1034	0.0194	0.0025	0.0002	0.0000	0.0000	0.0000	6
7		1.0000	0.9999	0.9992	0.9968	0.9906	0.9780	0.9562	0.9232	0.8779	0.5188	0.1904	0.0453	0.0073	0.0008	0.0001	0.0000	0.0000	7
8			1.0000	0.9999	0.9992	0.9973	0.9927	0.9833	0.9672	0.9421	0.6681	0.3073	0.0916	0.0183	0.0025	0.0002	0.0000	0.0000	8
9				1.0000	0.9998	0.9993	0.9978	0.9944	0.9875	0.9755	0.7911	0.4437	0.1637	0.0402	0.0067	0.0008	0.0001	0.0000	9
10					1.0000	0.9998	0.9994	0.9983	0.9957	0.9906	0.8801	0.5836	0.2622	0.0789	0.0160	0.0022	0.0002	0.0000	10
11						1.0000	0.9999	0.9995	0.9987	0.9968	0.9372	0.7107	0.3816	0.1390	0.0342	0.0057	0.0006	0.0000	11
12							1.0000	0.9999	0.9996	0.9990	0.9699	0.8139	0.5110	0.2229	0.0661	0.0133	0.0018	0.0002	12
13								1.0000	0.9999	0.9997	0.9868	0.8894	0.6370	0.3279	0.1163	0.0280	0.0045	0.0005	13
14									1.0000	0.9999	0.9947	0.9393	0.7481	0.4468	0.1878	0.0540	0.0104	0.0013	14
15										1.0000	0.9981	0.9692	0.8369	0.5692	0.2801	0.0955	0.0220	0.0033	15
16											0.9993	0.9856	0.9017	0.6839	0.3889	0.1561	0.0427	0.0077	16
17											0.9998	0.9937	0.9449	0.7822	0.5060	0.2369	0.0765	0.0164	17
18											0.9999	0.9975	0.9713	0.8594	0.6216	0.3356	0.1273	0.0325	18
19											1.0000	0.9991	0.9861	0.9152	0.7264	0.4465	0.1974	0.0595	19
20												0.9997	0.9937	0.9522	0.8139	0.5610	0.2862	0.1013	20
21												0.9999	0.9974	0.9749	0.8813	0.6701	0.3900	0.1611	21
22												1.0000	0.9990	0.9877	0.9290	0.7660	0.5019	0.2399	22
23													0.9996	0.9944	0.9604	0.8438	0.6134	0.3359	23
24													0.9999	0.9976	0.9793	0.9022	0.7160	0.4439	24
25													1.0000	0.9991	0.9900	0.9427	0.8034	0.5561	25
26														0.9997	0.9955	0.9686	0.8721	0.6641	26
27														0.9999	0.9981	0.9840	0.9220	0.7601	27
28														1.0000	0.9993	0.9924	0.9556	0.8389	28
29															0.9997	0.9966	0.9765	0.8987	29
30															0.9999	0.9986	0.9884	0.9405	30
31															1.0000	0.9995	0.9947	0.9675	31
32																0.9998	0.9978	0.9836	32
33																0.9999	0.9991	0.9923	33
34																1.0000	0.9997	0.9967	34
35																	0.9999	0.9987	35
36																	1.0000	0.9995	36
37																		0.9998	37
38																		1.0000	38

Table 2 Cumulative Poisson distribution function

The tabulated value is $P(X \leq x)$, where X has a Poisson distribution with mean λ.

x \ λ	0.1	0.2	0.3	0.4	0.5	0.6	0.7	0.8	0.9	1.0	1.2	1.4	1.6	1.8	λ \ x
0	0.9048	0.8187	0.7408	0.6703	0.6065	0.5488	0.4966	0.4493	0.4066	0.3679	0.3012	0.2466	0.2019	0.1653	0
1	0.9953	0.9825	0.9631	0.9384	0.9098	0.8781	0.8442	0.8088	0.7725	0.7358	0.6626	0.5918	0.5249	0.4628	1
2	0.9998	0.9989	0.9964	0.9921	0.9856	0.9769	0.9659	0.9526	0.9371	0.9197	0.8795	0.8335	0.7834	0.7306	2
3	1.0000	0.9999	0.9997	0.9992	0.9982	0.9966	0.9942	0.9909	0.9865	0.9810	0.9662	0.9463	0.9212	0.8913	3
4		1.0000	1.0000	0.9999	0.9998	0.9996	0.9992	0.9986	0.9977	0.9963	0.9923	0.9857	0.9763	0.9636	4
5				1.0000	1.0000	1.0000	0.9999	0.9998	0.9997	0.9994	0.9985	0.9968	0.9940	0.9896	5
6							1.0000	1.0000	1.0000	0.9999	0.9997	0.9994	0.9987	0.9974	6
7											1.0000	1.0000	0.9999	0.9997	0.9994	7
8													1.0000	1.0000	0.9999	8
9														1.0000	9

x \ λ	2.0	2.2	2.4	2.6	2.8	3.0	3.2	3.4	3.6	3.8	4.0	4.5	5.0	5.5	λ \ x
0	0.1353	0.1108	0.0907	0.0743	0.0608	0.0498	0.0408	0.0334	0.0273	0.0224	0.0183	0.0111	0.0067	0.0041	0
1	0.4060	0.3546	0.3084	0.2674	0.2311	0.1991	0.1712	0.1468	0.1257	0.1074	0.0916	0.0611	0.0404	0.0266	1
2	0.6767	0.6227	0.5697	0.5184	0.4695	0.4232	0.3799	0.3397	0.3027	0.2689	0.2381	0.1736	0.1247	0.0884	2
3	0.8571	0.8194	0.7787	0.7360	0.6919	0.6472	0.6025	0.5584	0.5152	0.4735	0.4335	0.3423	0.2650	0.2017	3
4	0.9473	0.9275	0.9041	0.8774	0.8477	0.8153	0.7806	0.7442	0.7064	0.6678	0.6288	0.5321	0.4405	0.3575	4
5	0.9834	0.9751	0.9643	0.9510	0.9349	0.9161	0.8946	0.8705	0.8441	0.8156	0.7851	0.7029	0.6160	0.5289	5
6	0.9955	0.9925	0.9884	0.9828	0.9756	0.9665	0.9554	0.9421	0.9267	0.9091	0.8893	0.8311	0.7622	0.6860	6
7	0.9989	0.9980	0.9967	0.9947	0.9919	0.9881	0.9832	0.9769	0.9692	0.9599	0.9489	0.9134	0.8666	0.8095	7
8	0.9998	0.9995	0.9991	0.9985	0.9976	0.9962	0.9943	0.9917	0.9883	0.9840	0.9786	0.9597	0.9319	0.8944	8
9	1.0000	0.9999	0.9998	0.9996	0.9993	0.9989	0.9982	0.9973	0.9960	0.9942	0.9919	0.9829	0.9682	0.9462	9
10		1.0000	1.0000	0.9999	0.9998	0.9997	0.9995	0.9992	0.9987	0.9981	0.9972	0.9933	0.9863	0.9747	10
11				1.0000	1.0000	0.9999	0.9999	0.9998	0.9996	0.9994	0.9991	0.9976	0.9945	0.9890	11
12						1.0000	1.0000	0.9999	0.9999	0.9998	0.9997	0.9992	0.9980	0.9955	12
13								1.0000	1.0000	1.0000	0.9999	0.9997	0.9993	0.9983	13
14											1.0000	0.9999	0.9998	0.9994	14
15												1.0000	0.9999	0.9998	15
16													1.0000	0.9999	16
17														1.0000	17

x \ λ	6.0	6.5	7.0	7.5	8.0	8.5	9.0	9.5	10.0	11.0	12.0	13.0	14.0	15.0	λ \ r
0	0.0025	0.0015	0.0009	0.0006	0.0003	0.0002	0.0001	0.0001	0.0000	0.0000	0.0000	0.0000	0.0000	0.0000	0
1	0.0174	0.0113	0.0073	0.0047	0.0030	0.0019	0.0012	0.0008	0.0005	0.0002	0.0001	0.0000	0.0000	0.0000	1
2	0.0620	0.0430	0.0296	0.0203	0.0138	0.0093	0.0062	0.0042	0.0028	0.0012	0.0005	0.0002	0.0001	0.0000	2
3	0.1512	0.1118	0.0818	0.0591	0.0424	0.0301	0.0212	0.0149	0.0103	0.0049	0.0023	0.0011	0.0005	0.0002	3
4	0.2851	0.2237	0.1730	0.1321	0.0996	0.0744	0.0550	0.0403	0.0293	0.0151	0.0076	0.0037	0.0018	0.0009	4
5	0.4457	0.3690	0.3007	0.2414	0.1912	0.1496	0.1157	0.0885	0.0671	0.0375	0.0203	0.0107	0.0055	0.0028	5
6	0.6063	0.5265	0.4497	0.3782	0.3134	0.2562	0.2068	0.1649	0.1301	0.0786	0.0458	0.0259	0.0142	0.0076	6
7	0.7440	0.6728	0.5987	0.5246	0.4530	0.3856	0.3239	0.2687	0.2202	0.1432	0.0895	0.0540	0.0316	0.0180	7
8	0.8472	0.7916	0.7291	0.6620	0.5925	0.5231	0.4557	0.3918	0.3328	0.2320	0.1550	0.0998	0.0621	0.0374	8
9	0.9161	0.8774	0.8305	0.7764	0.7166	0.6530	0.5874	0.5218	0.4579	0.3405	0.2424	0.1658	0.1094	0.0699	9
10	0.9574	0.9332	0.9015	0.8622	0.8159	0.7634	0.7060	0.6453	0.5830	0.4599	0.3472	0.2517	0.1757	0.1185	10
11	0.9799	0.9661	0.9467	0.9208	0.8881	0.8487	0.8030	0.7520	0.6968	0.5793	0.4616	0.3532	0.2600	0.1848	11
12	0.9912	0.9840	0.9730	0.9573	0.9362	0.9091	0.8758	0.8364	0.7916	0.6887	0.5760	0.4631	0.3585	0.2676	12
13	0.9964	0.9929	0.9872	0.9784	0.9658	0.9486	0.9261	0.8981	0.8645	0.7813	0.6815	0.5730	0.4644	0.3632	13
14	0.9986	0.9970	0.9943	0.9897	0.9827	0.9726	0.9585	0.9400	0.9165	0.8540	0.7720	0.6751	0.5704	0.4657	14
15	0.9995	0.9988	0.9976	0.9954	0.9918	0.9862	0.9780	0.9665	0.9513	0.9074	0.8444	0.7636	0.6694	0.5681	15
16	0.9998	0.9996	0.9990	0.9980	0.9963	0.9934	0.9889	0.9823	0.9730	0.9441	0.8987	0.8355	0.7559	0.6641	16
17	0.9999	0.9998	0.9996	0.9992	0.9984	0.9970	0.9947	0.9911	0.9857	0.9678	0.9370	0.8905	0.8272	0.7489	17
18	1.0000	0.9999	0.9999	0.9997	0.9993	0.9987	0.9976	0.9957	0.9928	0.9823	0.9626	0.9302	0.8826	0.8195	18
19		1.0000	1.0000	0.9999	0.9997	0.9995	0.9989	0.9980	0.9965	0.9907	0.9787	0.9573	0.9235	0.8752	19
20				1.0000	0.9999	0.9998	0.9996	0.9991	0.9984	0.9953	0.9884	0.9750	0.9521	0.9170	20
21					1.0000	0.9999	0.9998	0.9996	0.9993	0.9977	0.9939	0.9859	0.9712	0.9469	21
22						1.0000	0.9999	0.9999	0.9997	0.9990	0.9970	0.9924	0.9833	0.9673	22
23							1.0000	0.9999	0.9999	0.9995	0.9985	0.9960	0.9907	0.9805	23
24								1.0000	1.0000	0.9998	0.9993	0.9980	0.9950	0.9888	24
25										0.9999	0.9997	0.9990	0.9974	0.9938	25
26										1.0000	0.9999	0.9995	0.9987	0.9967	26
27											0.9999	0.9998	0.9994	0.9983	27
28											1.0000	0.9999	0.9997	0.9991	28
29												1.0000	0.9999	0.9996	29
30													0.9999	0.9998	30
31													1.0000	0.9999	31
32														1.0000	32

Table 3 Normal distribution function

The table gives the probability p that a normally distributed random variable Z, with mean = 0 and variance = 1, is less than or equal to z.

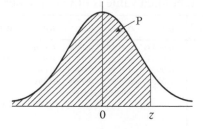

z	0.00	0.01	0.02	0.03	0.04	0.05	0.06	0.07	0.08	0.09	z
0.1	0.50000	0.50399	0.50798	0.51197	0.51595	0.51994	0.52392	0.52790	0.53188	0.53586	0.0
0.1	0.53983	0.54380	0.54776	0.55172	0.55567	0.55962	0.56356	0.56749	0.57142	0.57535	0.1
0.2	0.57926	0.58317	0.58706	0.59095	0.59483	0.59871	0.60257	0.60642	0.61026	0.61409	0.2
0.3	0.61791	0.62172	0.62552	0.62930	0.63307	0.63683	0.64058	0.64431	0.64803	0.65173	0.3
0.4	0.65542	0.65910	0.66276	0.66640	0.67003	0.67364	0.67724	0.68082	0.68439	0.68793	0.4
0.5	0.69146	0.69497	0.69847	0.70194	0.70540	0.70884	0.71226	0.71566	0.71904	0.72240	0.5
0.6	0.72575	0.72907	0.73237	0.73565	0.73891	0.74215	0.74537	0.74857	0.75175	0.75490	0.6
0.7	0.75804	0.76115	0.76424	0.76730	0.77035	0.77337	0.77637	0.77935	0.78230	0.78524	0.7
0.8	0.78814	0.79103	0.79389	0.79673	0.79955	0.80234	0.80511	0.80785	0.81057	0.81327	0.8
0.9	0.81594	0.81859	0.82121	0.82381	0.82639	0.82894	0.83147	0.83398	0.83646	0.83891	0.9
1.0	0.84134	0.84375	0.84614	0.84849	0.85083	0.85314	0.85543	0.85769	0.85993	0.86214	1.0
1.1	0.86433	0.86650	0.86864	0.87076	0.87286	0.87493	0.87698	0.87900	0.88100	0.88298	1.1
1.2	0.88493	0.88686	0.88877	0.89065	0.89251	0.89435	0.89617	0.89796	0.89973	0.90147	1.2
1.3	0.90320	0.90490	0.90658	0.90824	0.90988	0.91149	0.91309	0.91466	0.91621	0.91774	1.3
1.4	0.91924	0.92073	0.92220	0.92364	0.92507	0.92647	0.92785	0.92922	0.93056	0.93189	1.4
1.5	0.93319	0.93448	0.93574	0.93699	0.93822	0.93943	0.94062	0.94179	0.94295	0.94408	1.5
1.6	0.94520	0.94630	0.94738	0.94845	0.94950	0.95053	0.95154	0.95254	0.95352	0.95449	1.6
1.7	0.95543	0.95637	0.95728	0.95818	0.95907	0.95994	0.96080	0.96164	0.96246	0.96327	1.7
1.8	0.96407	0.96485	0.96562	0.96638	0.96712	0.96784	0.96856	0.96926	0.96995	0.97062	1.8
1.9	0.97128	0.97193	0.97257	0.97320	0.97381	0.97441	0.97500	0.97558	0.97615	0.97670	1.9
2.0	0.97725	0.97778	0.97831	0.97882	0.97932	0.97982	0.98030	0.98077	0.98124	0.98169	2.0
2.1	0.98214	0.98257	0.98300	0.98341	0.98382	0.98422	0.98461	0.98500	0.98537	0.98574	2.1
2.2	0.98610	0.98645	0.98679	0.98713	0.98745	0.98778	0.98809	0.98840	0.98870	0.98899	2.2
2.3	0.98928	0.98956	0.98983	0.99010	0.99036	0.99061	0.99086	0.99111	0.99134	0.99158	2.3
2.4	0.99180	0.99202	0.99224	0.99245	0.99266	0.99286	0.99305	0.99324	0.99343	0.99361	2.4
2.5	0.99379	0.99396	0.99413	0.99430	0.99446	0.99461	0.99477	0.99492	0.99506	0.99520	2.5
2.6	0.99534	0.99547	0.99560	0.99573	0.99585	0.99598	0.99609	0.99621	0.99632	0.99643	2.6
2.7	0.99653	0.99664	0.99674	0.99683	0.99693	0.99702	0.99711	0.99720	0.99728	0.99736	2.7
2.8	0.99744	0.99752	0.99760	0.99767	0.99774	0.99781	0.99788	0.99795	0.99801	0.99807	2.8
2.9	0.99813	0.99819	0.99825	0.99831	0.99836	0.99841	0.99846	0.99851	0.99856	0.99861	2.9
3.0	0.99865	0.99869	0.99874	0.99878	0.99882	0.99886	0.99889	0.99893	0.99896	0.99900	3.0
3.1	0.99903	0.99906	0.99910	0.99913	0.99916	0.99918	0.99921	0.99924	0.99926	0.99929	3.1
3.2	0.99931	0.99934	0.99936	0.99938	0.99940	0.99942	0.99944	0.99946	0.99948	0.99950	3.2
3.3	0.99952	0.99953	0.99955	0.99957	0.99958	0.99960	0.99961	0.99962	0.99964	0.99965	3.3
3.4	0.99966	0.99968	0.99969	0.99970	0.99971	0.99972	0.99973	0.99974	0.99975	0.99976	3.4
3.5	0.99977	0.99978	0.99978	0.99979	0.99980	0.99981	0.99981	0.99982	0.99983	0.99983	3.5
3.6	0.99984	0.99985	0.99985	0.99986	0.99986	0.99987	0.99987	0.99988	0.99988	0.99989	3.6
3.7	0.99989	0.99990	0.99990	0.99990	0.99991	0.99991	0.99992	0.99992	0.99992	0.99992	3.7
3.8	0.99993	0.99993	0.99993	0.99994	0.99994	0.99994	0.99994	0.99995	0.99995	0.99995	3.8
3.9	0.99995	0.99995	0.99996	0.99996	0.99996	0.99996	0.99996	0.99996	0.99997	0.99997	3.9

Table 4 Percentage points of the normal distribution

The table gives the values of z satisfying $P(Z \leqslant z) = p$,
where Z is the normally distributed random variable with
mean $= 0$ and variance $= 1$.

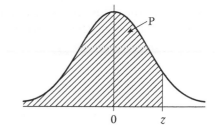

p	0.00	0.01	0.02	0.03	0.04	0.05	0.06	0.07	0.08	0.09	p
0.5	0.0000	0.0251	0.0502	0.0753	0.1004	0.1257	0.1510	0.1764	0.2019	0.2275	**0.5**
0.6	0.2533	0.2793	0.3055	0.3319	0.3585	0.3853	0.4125	0.4399	0.4677	0.4958	**0.6**
0.7	0.5244	0.5534	0.5828	0.6128	0.6433	0.6745	0.7063	0.7388	0.7722	0.8064	**0.7**
0.8	0.8416	0.8779	0.9154	0.9542	0.9945	1.0364	1.0803	1.1264	1.1750	1.2265	**0.8**
0.9	1.2816	1.3408	1.4051	1.4758	1.5548	1.6449	1.7507	1.8808	2.0537	2.3263	**0.9**

p	0.000	0.001	0.002	0.003	0.004	0.005	0.006	0.007	0.008	0.009	p
0.95	1.6449	1.6546	1.6646	1.6747	1.6849	1.6954	1.7060	1.7169	1.7279	1.7392	**0.95**
0.96	1.7507	1.7624	1.7744	1.7866	1.7991	1.8119	1.8250	1.8384	1.8522	1.8663	**0.96**
0.97	1.8808	1.8957	1.9110	1.9268	1.9431	1.9600	1.9774	1.9954	2.0141	2.0335	**0.97**
0.98	2.0537	2.0749	2.0969	2.1201	2.1444	2.1701	2.1973	2.2262	2.2571	2.2904	**0.98**
0.99	2.3263	2.3656	2.4089	2.4573	2.5121	2.5758	2.6521	2.7478	2.8782	3.0902	**0.99**

Table 6 Percentage points of the χ^2 distribution

The table gives the values of x satisfying $P(X \leqslant X) = p$, where X is a random variable having the χ^2 distribution with ν degrees of freedom.

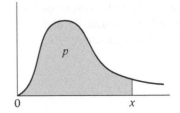

ν \ p	0.005	0.01	0.025	0.05	0.1	0.9	0.95	0.975	0.99	0.995	ν \ p
1	0.00004	0.0002	0.001	0.004	0.016	2.706	3.841	5.024	6.635	7.879	1
2	0.010	0.020	0.051	0.103	0.211	4.605	5.991	7.378	9.210	10.597	2
3	0.072	0.115	0.216	0.352	0.584	6.251	7.815	9.348	11.345	12.838	3
4	0.207	0.297	0.484	0.711	1.064	7.779	9.488	11.143	13.277	14.860	4
5	0.412	0.554	0.831	1.145	1.610	9.236	11.070	12.833	15.086	16.750	5
6	0.676	0.872	1.237	1.635	2.204	10.645	12.592	14.449	16.812	18.548	6
7	0.989	1.239	1.690	2.167	2.833	12.017	14.067	16.013	18.475	20.278	7
8	1.344	1.646	2.180	2.733	3.490	13.362	15.507	17.535	20.090	21.955	8
9	1.735	2.088	2.700	3.325	4.168	14.684	16.919	19.023	21.666	23.589	9
10	2.156	2.558	3.247	3.940	4.865	15.987	18.307	20.483	23.209	25.188	10
11	2.603	3.053	3.816	4.575	5.578	17.275	19.675	21.920	24.725	26.757	11
12	3.074	3.571	4.404	5.226	6.304	18.549	21.026	23.337	26.217	28.300	12
13	3.565	4.107	5.009	5.892	7.042	19.812	22.362	24.736	27.688	29.819	13
14	4.075	4.660	5.629	6.571	7.790	21.064	23.685	26.119	29.141	31.319	14
15	4.601	5.229	6.262	7.261	8.547	22.307	24.996	27.488	30.578	32.801	15
16	5.142	5.812	6.908	7.962	9.312	23.542	26.296	28.845	32.000	34.267	16
17	5.697	6.408	7.564	8.672	10.085	24.769	27.587	30.191	33.409	35.718	17
18	6.265	7.015	8.231	9.390	10.865	25.989	28.869	31.526	34.805	37.156	18
19	6.844	7.633	8.907	10.117	11.651	27.204	30.144	32.852	36.191	38.582	19
20	7.434	8.260	9.591	10.851	12.443	28.412	31.410	34.170	37.566	39.997	20
21	8.034	8.897	10.283	11.591	13.240	29.615	32.671	35.479	38.932	41.401	21
22	8.643	9.542	10.982	12.338	14.041	30.813	33.924	36.781	40.289	42.796	22
23	9.260	10.196	11.689	13.091	14.848	32.007	35.172	38.076	41.638	44.181	23
24	9.886	10.856	12.401	13.848	15.659	33.196	36.415	39.364	42.980	45.559	24
25	10.520	11.524	13.120	14.611	16.473	34.382	37.652	40.646	44.314	46.928	25
26	11.160	12.198	13.844	15.379	17.292	35.563	38.885	41.923	45.642	48.290	26
27	11.808	12.879	14.573	16.151	18.114	36.741	40.113	43.195	46.963	49.645	27
28	12.461	13.565	15.308	16.928	18.939	37.916	41.337	44.461	48.278	50.993	28
29	13.121	14.256	16.047	17.708	19.768	39.087	42.557	45.722	49.588	52.336	29
30	13.787	14.953	16.791	18.493	20.599	40.256	43.773	46.979	50.892	53.672	30
31	14.458	15.655	17.539	19.281	21.434	41.422	44.985	48.232	52.191	55.003	31
32	15.134	16.362	18.291	20.072	22.271	42.585	46.194	49.480	53.486	56.328	32
33	15.815	17.074	19.047	20.867	23.110	43.745	47.400	50.725	54.776	57.648	33
34	16.501	17.789	19.806	21.664	23.952	44.903	48.602	51.996	56.061	58.964	34
35	17.192	18.509	20.569	22.465	24.797	46.059	49.802	53.203	57.342	60.275	35
36	17.887	19.223	21.336	23.269	25.643	47.212	50.998	54.437	58.619	61.581	36
37	18.586	19.960	22.106	24.075	26.492	48.363	52.192	55.668	59.892	62.883	37
38	19.289	20.691	22.878	24.884	27.343	49.513	53.384	56.896	61.162	64.181	38
39	19.996	21.426	23.654	25.695	28.196	50.660	54.572	58.120	62.428	65.476	39
40	20.707	22.164	24.433	26.509	29.051	51.805	55.758	59.342	63.691	66.766	40
45	24.311	25.901	28.366	30.612	33.350	57.505	61.656	65.410	69.957	73.166	45
50	27.991	29.707	32.357	34.764	37.689	63.167	67.505	71.420	76.154	79.490	50
55	31.735	33.570	36.398	38.958	42.060	68.796	73.311	77.380	82.292	85.749	55
60	35.534	37.485	40.482	43.188	46.459	74.397	79.082	83.298	88.379	91.952	60
65	39.383	41.444	44.603	47.450	50.883	79.973	84.821	89.177	94.422	98.105	65
70	43.275	45.442	48.758	51.739	55.329	85.527	90.531	95.023	100.425	104.215	70
75	47.206	49.475	52.942	56.054	59.795	91.061	96.217	100.839	106.393	110.286	75
80	51.172	53.540	57.153	60.391	64.278	96.578	101.879	106.629	112.329	116.321	80
85	55.170	57.634	61.389	64.749	68.777	102.079	107.522	112.393	118.236	122.325	85
90	59.196	61.754	65.647	69.126	73.291	107.565	113.145	118.136	124.116	128.299	90
95	63.250	65.898	69.925	73.520	77.818	113.038	118.752	123.858	129.973	134.247	95
100	67.328	70.065	74.222	77.929	82.358	118.498	124.342	129.561	135.807	140.169	100

Table 8 Critical values of the product moment correlation coefficient

The table gives the critical values, for different significance levels, of the product moment correlation coefficient, *r*, for varying sample sizes, *n*.

One tail Two tail	10% 20%	5% 10%	2.5% 5%	1% 2%	0.5% 1%	One tail Two tail
n						*n*
4	0.8000	0.9000	0.9500	0.9800	0.9900	4
5	0.6870	0.8054	0.8783	0.9343	0.9587	5
6	0.6084	0.7293	0.8114	0.8822	0.9172	6
7	0.5509	0.6694	0.7545	0.8329	0.8745	7
8	0.5067	0.6215	0.7067	0.7887	0.8343	8
9	0.4716	0.5822	0.6664	0.7498	0.7977	9
10	0.4428	0.5494	0.6319	0.7155	0.7646	10
11	0.4187	0.5214	0.6021	0.6851	0.7348	11
12	0.3981	0.4973	0.5760	0.6581	0.7079	12
13	0.3802	0.4762	0.5529	0.6339	0.6835	13
14	0.3646	0.4575	0.5324	0.6120	0.6614	14
15	0.3507	0.4409	0.5140	0.5923	0.6411	15
16	0.3383	0.4259	0.4973	0.5742	0.6226	16
17	0.3271	0.4124	0.4821	0.5577	0.6055	17
18	0.3170	0.4000	0.4683	0.5425	0.5897	18
19	0.3077	0.3887	0.4555	0.5285	0.5751	19
20	0.2992	0.3783	0.4438	0.5155	0.5614	20
21	0.2914	0.3687	0.4329	0.5034	0.5487	21
22	0.2841	0.3598	0.4227	0.4921	0.5368	22
23	0.2774	0.3515	0.4132	0.4815	0.5256	23
24	0.2711	0.3438	0.4044	0.4716	0.5151	24
25	0.2653	0.3365	0.3961	0.4622	0.5052	25
26	0.2598	0.3297	0.3882	0.4534	0.4958	26
27	0.2546	0.3233	0.3809	0.4451	0.4869	27
28	0.2497	0.3172	0.3739	0.4372	0.4785	28
29	0.2451	0.3115	0.3673	0.4297	0.4705	29
30	0.2407	0.3061	0.3610	0.4226	0.4629	30
31	0.2366	0.3009	0.3550	0.4158	0.4556	31
32	0.2327	0.2960	0.3494	0.4093	0.4487	32
33	0.2289	0.2913	0.3440	0.4032	0.4421	33
34	0.2254	0.2869	0.3388	0.3972	0.4357	34
35	0.2220	0.2826	0.3338	0.3916	0.4296	35
36	0.2187	0.2785	0.3291	0.3862	0.4238	36
37	0.2156	0.2746	0.3246	0.3810	0.4182	37
38	0.2126	0.2709	0.3202	0.3760	0.4128	38
39	0.2097	0.2673	0.3160	0.3712	0.4076	39
40	0.2070	0.2638	0.3120	0.3665	0.4026	40
41	0.2043	0.2605	0.3081	0.3621	0.3978	41
42	0.2018	0.2573	0.3044	0.3578	0.3932	42
43	0.1993	0.2542	0.3008	0.3536	0.3887	43
44	0.1970	0.2512	0.2973	0.3496	0.3843	44
45	0.1947	0.2483	0.2940	0.3457	0.3801	45
46	0.1925	0.2455	0.2907	0.3420	0.3761	46
47	0.1903	0.2429	0.2876	0.3384	0.3721	47
48	0.1883	0.2403	0.2845	0.3348	0.3683	48
49	0.1863	0.2377	0.2816	0.3314	0.3646	49
50	0.1843	0.2353	0.2787	0.3281	0.3610	50
60	0.1678	0.2144	0.2542	0.2997	0.3301	60
70	0.1550	0.1982	0.2352	0.2776	0.3060	70
80	0.1448	0.1852	0.2199	0.2597	0.2864	80
90	0.1364	0.1745	0.2072	0.2449	0.2702	90
100	0.1292	0.1654	0.1966	0.2324	0.2565	100

Table 9 Critical values of Spearman's rank correlation coefficient

The table gives the critical values, for different significance levels, of Spearman's rank correlation coefficient, r_s, for varying sample sizes, n.

Since r_s is discrete, exact significance levels cannot be obtained in most cases.

The critical values given are those with significance levels closest to the stated value.

One tail Two tail	10% 20%	5% 10%	2.5% 5%	1% 2%	0.5% 1%	One tail Two tail
n						n
4	1.0000	1.0000	1.0000	1.0000	1.0000	4
5	0.7000	0.9000	0.9000	1.0000	1.0000	5
6	0.6571	0.7714	0.8286	0.9429	0.9429	6
7	0.5714	0.6786	0.7857	0.8571	0.8929	7
8	0.5476	0.6429	0.7381	0.8095	0.8571	8
9	0.4833	0.6000	0.6833	0.7667	0.8167	9
10	0.4424	0.5636	0.6485	0.7333	0.7818	10
11	0.4182	0.5273	0.6091	0.7000	0.7545	11
12	0.3986	0.5035	0.5874	0.6713	0.7273	12
13	0.3791	0.4780	0.5604	0.6484	0.6978	13
14	0.3670	0.4593	0.5385	0.6220	0.6747	14
15	0.3500	0.4429	0.5179	0.6000	0.6536	15
16	0.3382	0.4265	0.5029	0.5824	0.6324	16
17	0.3271	0.4124	0.4821	0.5577	0.6055	17
18	0.3170	0.4000	0.4683	0.5425	0.5897	18
19	0.3077	0.3887	0.4555	0.5285	0.5751	19
20	0.2992	0.3783	0.4438	0.5155	0.5614	20
21	0.2914	0.3687	0.4329	0.5034	0.5487	21
22	0.2841	0.3598	0.4227	0.4921	0.5368	22
23	0.2774	0.3515	0.4132	0.4815	0.5256	23
24	0.2711	0.3438	0.4044	0.4716	0.5151	24
25	0.2653	0.3365	0.3961	0.4622	0.5052	25
26	0.2598	0.3297	0.3882	0.4534	0.4958	26
27	0.2546	0.3233	0.3809	0.4451	0.4869	27
28	0.2497	0.3172	0.3739	0.4372	0.4785	28
29	0.2451	0.3115	0.3673	0.4297	0.4705	29
30	0.2407	0.3061	0.3610	0.4226	0.4629	30
31	0.2366	0.3009	0.3550	0.4158	0.4556	31
32	0.2327	0.2960	0.3494	0.4093	0.4487	32
33	0.2289	0.2913	0.3440	0.4032	0.4421	33
34	0.2254	0.2869	0.3388	0.3972	0.4357	34
35	0.2220	0.2826	0.3338	0.3916	0.4296	35
36	0.2187	0.2785	0.3291	0.3862	0.4238	36
37	0.2156	0.2746	0.3246	0.3810	0.4182	37
38	0.2126	0.2709	0.3202	0.3760	0.4128	38
39	0.2097	0.2673	0.3160	0.3712	0.4076	39
40	0.2070	0.2638	0.3120	0.3665	0.4026	40
41	0.2043	0.2605	0.3081	0.3621	0.3978	41
42	0.2018	0.2573	0.3044	0.3578	0.3932	42
43	0.1993	0.2542	0.3008	0.3536	0.3887	43
44	0.1970	0.2512	0.2973	0.3496	0.3843	44
45	0.1947	0.2483	0.2940	0.3457	0.3801	45
46	0.1925	0.2455	0.2907	0.3420	0.3761	46
47	0.1903	0.2429	0.2876	0.3384	0.3721	47
48	0.1883	0.2403	0.2845	0.3348	0.3683	48
49	0.1863	0.2377	0.2816	0.3314	0.3646	49
50	0.1843	0.2353	0.2787	0.3281	0.3610	50
60	0.1678	0.2144	0.2542	0.2997	0.3301	60
70	0.1550	0.1982	0.2352	0.2776	0.3060	70
80	0.1448	0.1852	0.2199	0.2597	0.2864	80
90	0.1364	0.1745	0.2072	0.2449	0.2702	90
100	0.1292	0.1654	0.1966	0.2324	0.2565	100

Table 10 Critical values of the Wilcoxon signed-rank statistic

The table gives the lower tail critical values of the statistic T.

The upper tail critical values are given by $\frac{1}{2}n(n+1) - T$.

T is the sum of the ranks of observations with the same sign.
Since T is discrete, exact significance levels cannot usually be obtained.
The critical values tabulated are those with significance levels closest to the stated value.
The critical region includes the tabulated value.

One tail Two tail	10% 20%	5% 10%	2.5% 5%	1% 2%	0.5% 1%
n					
3	0				
4	1	0			
5	2	1	0		
6	4	2	1	0	
7	6	4	2	0	0
8	8	6	4	2	0
9	11	8	6	3	2
10	14	11	8	5	3
11	18	14	11	7	5
12	22	17	14	10	7
13	26	21	17	13	10
14	31	26	21	16	13
15	37	30	25	20	16
16	42	36	30	24	19
17	49	41	35	28	23
18	55	47	40	33	28
19	62	54	46	38	32
20	70	60	52	43	37

Table 11 Critical values of the Mann–Whitney U statistic

The table gives the lower tail critical values of the statistic U.

The upper tail critical values are given by $mn - U$.

$U = T - \dfrac{n(n+1)}{2}$ where T is the sum of the ranks of the sample of size n.

Since T is discrete, exact significance levels cannot be obtained.
The critical values tabulated are those with significance levels closest to the stated value.
The critical region includes the tabulated value.

One tail 5% Two tail 10% n / m	2	3	4	5	6	7	8	9	10	11	12
2		0	0	0	0	1	1	1	2	2	2
3	0	0	1	1	2	3	3	4	5	5	6
4	0	1	2	3	4	5	6	7	8	9	10
5	0	1	3	4	5	7	8	10	11	12	14
6	0	2	4	5	7	9	11	12	14	16	18
7	1	3	5	7	9	11	13	15	18	20	22
8	1	3	6	8	11	13	16	18	21	24	26
9	1	4	7	10	12	15	18	21	24	27	30
10	2	5	8	11	14	18	21	24	28	31	34
11	2	5	9	12	16	20	24	27	31	35	39
12	2	6	10	14	18	22	26	30	34	39	43

One tail 2.5% Two tail 5% n / m	2	3	4	5	6	7	8	9	10	11	12
2				0	0	0	0	0	1	1	1
3			0	0	1	2	2	3	3	4	4
4		0	1	2	2	3	4	5	6	7	8
5	0	0	2	3	4	5	6	7	9	10	11
6	0	1	2	4	5	7	8	10	12	13	15
7	0	2	3	5	7	9	11	13	15	17	18
8	0	2	4	6	8	11	13	15	18	20	22
9	0	3	5	7	10	13	15	18	21	23	26
10	1	3	6	9	12	15	18	21	24	27	30
11	1	4	7	10	13	17	20	23	27	30	34
12	1	4	8	11	15	18	22	26	30	34	38

Answers

EXERCISE 1A

1 (a) (i)

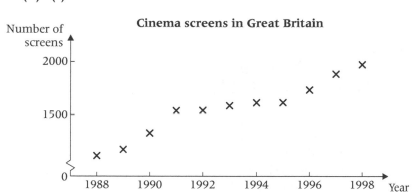

The trend is upward and approximately linear.

(ii)

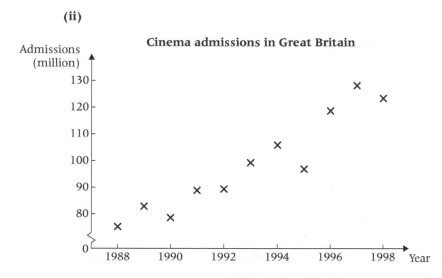

The trend is upward and approximately linear.

(b) Both trends are upward and approximately linear but, compared to the number of screens, the admissions show more variability about the trend.

2

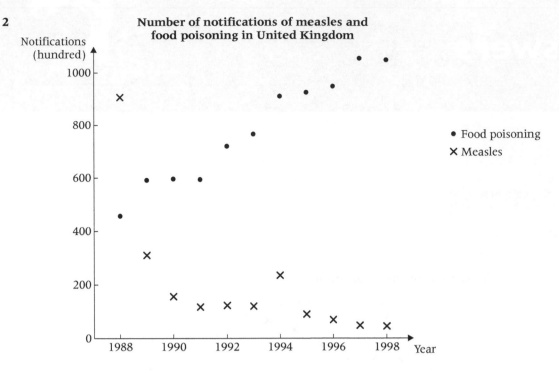

Number of notifications of measles and food poisoning in United Kingdom

Note: as notifications of measles and of food poisoning were both in hundreds it was possible to show them both on the same graph. This is not essential but may be helpful when making comparisons.

(a) Notifications of measles showed a downward non-linear trend. The drop was particularly sharp up to 1990. There were approximately double the number of cases in 1994 than would have been expected from the trend.

(b) Notifications of food poisoning shows an upward, approximately linear trend.

3 (a)

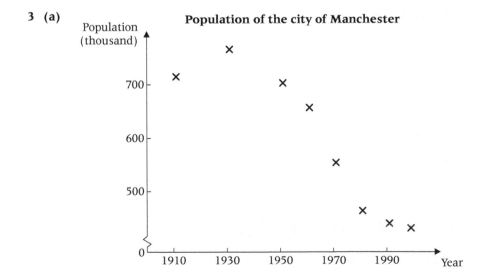

Population of the city of Manchester

Note. Unusually for a time series the points are not evenly spaced along the horizontal axis.

From 1930 the trend is downward but not linear.

(b)

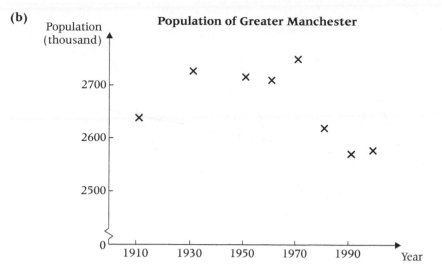

Note: it would have been possible to put both series on the same graph. This would have had advantages for making comparisons but it would have been difficult to see the trend in the City population as this is much smaller than the population of Greater Manchester.

There has been some reduction in the population since 1930 but there is no clear trend in this data.

The reduction in the population of the City has been much greater than the reduction in the population of Greater Manchester. The proportion of the population of greater Manchester living in the City has reduced from more than a quarter in 1911 to about a sixth in 1998.

EXERCISE 1B

1

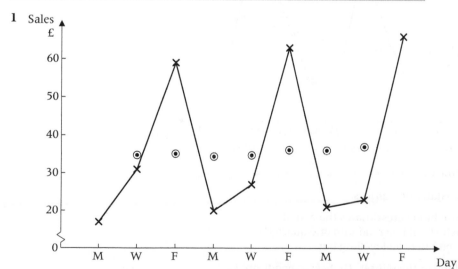

2 (a)

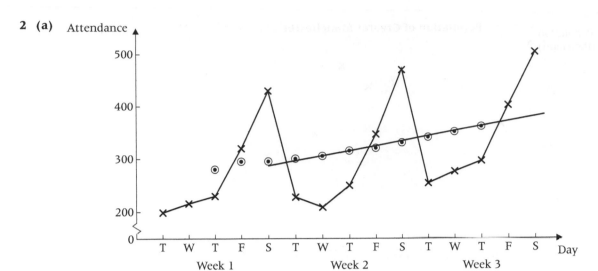

(b) Wednesday, week 2. Attendance was substantially further below the moving average than the other two Wednesdays.

(c) Approximately 290.

(d) The moving average shows an upward approximately linear trend. This cannot continue in the long term since the theatre holds a maximum of 600 people.

3 (a) 152.25 (£152 250); **(b)** 176.3 (£176 300).

4 (a)

Expenditure of UK households on horticultural goods (1995 prices)

(b) Upward, approximately linear.

(c) About £810 000 000 at constant 1995 prices.

(d) Forecast is reasonably close but overestimates the actual expenditure. Not too much should be read into one quarter's figures but the upward trend may be levelling out.

(e) Not much gardening is done in the winter. Highest expenditure is in Quarter 2 when people are preparing to resume.

5 (a)

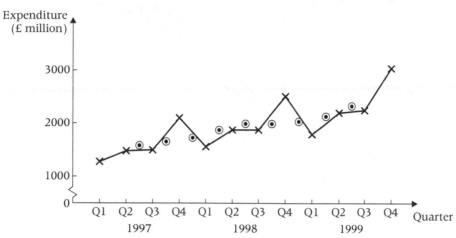

Expenditure of UK households on sports and toys (1995 prices)

(b) Upward, approximately linear.

(c) Predicted moving averages for 2000 in £million, 1995 prices:

Q1 – 2510 Q2 – 2600 Q3 – 2690 Q4 – 2780.

(Note: if $t = 1$ for Q1, 1997, then first moving average corresponds to $t = 2.5$.)

(d) Predicted expenditure 2000 in £million, 1995 prices, estimated to nearest 50:

Q1 – 2200 Q2 – 2550 Q3 – 2550 Q4 – 3200.

(e) Forecast is reasonably close but overestimates the actual expenditure. Not too much should be read into a single quarter but the rate of increase may be reducing.

(f) Highest expenditure is in Quarter 4 which includes Christmas.

EXERCISE 1C

1 Possible answers

(a)

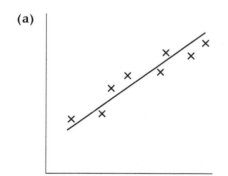

(b)

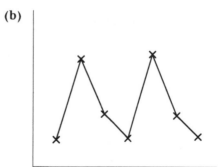

(c)

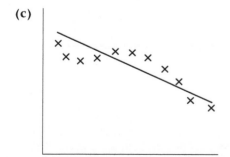

(d)

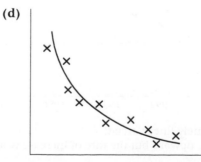

(e)

(f)

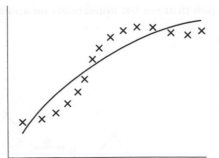

2 (a) A; **(b)** E; **(c)** F; **(d)** B.

3 (a)

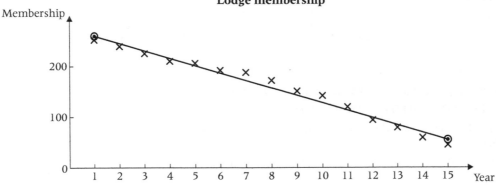

Lodge membership

(b) $y = 276 - 14.8t$;
(c) Short-term variability about a downward, linear trend;
(d) 40;
(e) Graph suggests the actual the value will be below the regression line, say 30;
(f) The regression line would predict a negative membership for 2005 which is impossible.

MIXED EXERCISE

1 (a)

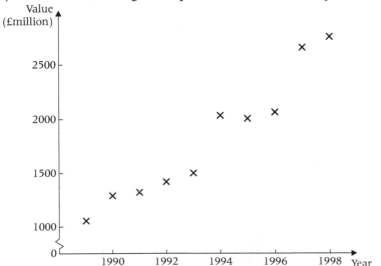

(b) Upward, approximately linear, trend;
(c) Trend would still be upward but the rate of increase would be less.

2 (a)

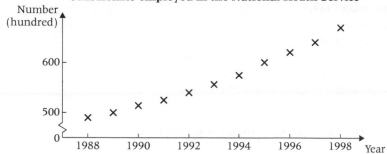

Consultants employed in the National Heath Service

Number of consultants shows an approximately linear upward trend.

(b)

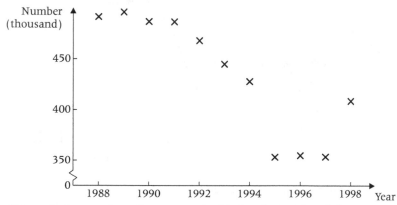

Nurses and midwives employed in the National Heath Service

Although the number of nurses and midwives has reduced there is no consistent trend to the data.

In 1988 there were about ten times as many nurses and midwives employed in the National Health Service as there were consultants. The number of consultants has shown a steady upward trend while the number of nurses and midwives has decreased. In 1998 there were only about six times as many nurses and midwives as consultants.

3 (a)

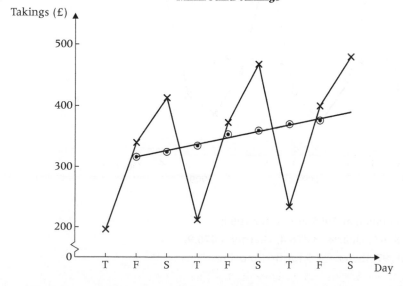

Milkround takings

(b) Predicted moving average for next week:
Tuesday £395, Friday £405, Saturday £415;

(c) Predicted takings for next week:
Tuesday £270, Friday £430, Saturday £520;

(d) It would be foolish to use this small amount of data to predict takings
a year ahead. The trend is unlikely to continue unchanged and even if
it did this amount of data will give only a rough estimate of the trend.
Projecting this estimate a year ahead could lead to major errors.

4 (a)

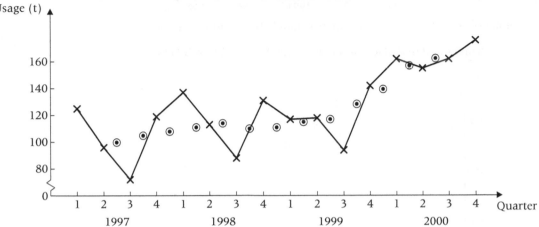

Oil usage of small engineering firm

(b) Quarter 1, 1999, usage is close to moving average. For all other first
quarters the usage is well above the moving average;

(c) Quarter 3, 2000, usage is close to moving average. For other third
quarters the usage is well below the moving average.

5 (a)

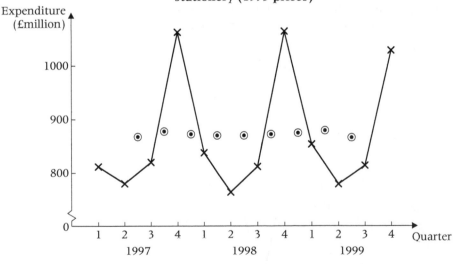

**Expenditure of UK households on
stationery (1995 prices)**

(b) Predicted moving average, £million at 1995 prices, for 2000:

Quarter 1 875.5, Quarter 2 876.0, Quarter 3 876.4, Quarter 4 876.9.

Note: if $t = 1$ for quarter 1 1997, then $t = 2.5$ for first moving
average point.

(c) Predicted actual expenditure, £million at 1995 prices, for 2000
Quarter 1 840, Quarter 2 780, Quarter 3 820, Quarter 4 1060;

(d) Prediction was a slight underestimate but fairly accurate. Method of forecasting seems satisfactory in this case.

6 (a)

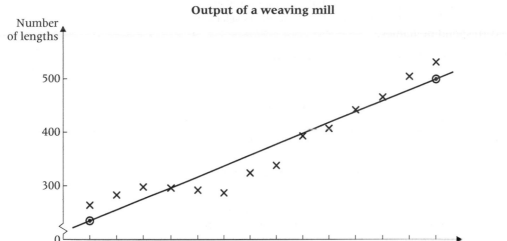

Output of a weaving mill

(b) $y = 212 + 20.5t$;

(c) Short-term variability about an upward linear trend;

(d) 520;

(e) 550.

2 Sampling

EXERCISE 2A

1 Number books 0000 to 2124. Select four-digit random numbers. Ignore repeats and >2124. Continue until 20 numbers obtained. Select corresponding books.

2 Obtain list of employees names and number 000 to 711.
Select three-digit random numbers. Ignore repeats and >711. Continue until six numbers obtained. Select corresponding employees.

3 Number the points 00 to 27. Select two-digit random numbers. Ignore repeats and >27. Continue until eight numbers obtained. Select corresponding plants.

4 (a) There are 36 students ages. Allocate the number 00 to the first age (19), the number 01 to the second age (20) ... allocate the number 35 to the last age (27).

Select two-digit random numbers ignoring >35

62 50 62 27 80 30 72 07 93 38 68 35 86 27 65 33
       27   38   19   27   27   41

The ages selected are 27, 38, 19, 27, 27, 41. (Answers will vary according to your starting point in the random number tables.)

(b) As in part (a) except that now repeats of random numbers are ignored.

62 50 62 27 80 30 72 07 93 38 68 35 86 27 65 33 40 18
       27   38   19   27   41   20

The ages selected are 27, 38, 19, 27, 41, 20.

5 Number rods 000 to 499. Select 3-digit random numbers. Ignore repeats and >499. Continue until 20 numbers selected. Choose corresponding rods.

6 (a) No, because not all sets of 40 customers can be chosen.

(b) Number names 0000 to 8949. Select 4-digit random numbers. Ignore repeats and >8949. Continue until 40 numbers selected. Choose corresponding names.

7 No, because not all sets of 128 electors can be selected.

EXERCISE 2B

1 Prior knowledge is used to divide the population into strata. Samples are taken from each of the strata in proportion to the size of each of the strata.

Advantage – ensures sample contains same proportion of each of the strata as the population (unlike a random sample).

Disadvantage – more complicated to obtain (or needs relevant prior knowledge).

2 Number males 00 to 11; choose two-digit random numbers; ignore repeats and >11; continue until four obtained; select corresponding males. Choose six of the 18 females in the same way.

3 (a) (i) Stratified sampling will ensure that the sample contains men and women in the same proportion as the population. If there is a difference in attitudes between genders this will ensure the population is fairly represented.

(ii) The additional information needed and extra difficulty of obtaining a stratified sample will be to no purpose if there is no difference in attitude between genders.

(b) age, parent/not parent, employed/unemployed, etc.

EXERCISE 2C

1 Prior knowledge of the population is used to divide it into strata. Samples from each of the strata (usually in proportion to the size of the strata) are taken by any convenient method. Use when relevant prior knowledge is available and random sampling is too difficult or too expensive.

2 Please distribute the questionnaire to nine sixth-form girls, seven sixth-form boys, 19 girls who are not in the sixth form and 20 boys who are not in the sixth form. Provided you stick to these numbers you may ask the first pupils you see or use any other convenient method to select the sample.

3 (a) (i) C, **(ii)** A random, B stratified;

(b) A equally likely, B equally likely, C not equally likely (or depends on the 'convenient method' chosen);

(c) (i) no possibility of bias,

(ii) easier to carry out (also ensures all factories fairly represented);

(d) Both avoid the possibility of bias, but B ensures all factories are fairly represented.

EXERCISE 2D

1 Population is divided into clusters (usually geographically). A sample of clusters is selected at random and then all, or a random sample of, individuals within the chosen clusters form the sample. For example, to choose a sample of oil tanker drivers in UK, choose a random sample of oil distribution depots and then a random sample of drivers from the chosen depots. This sample will be more geographically localised than a random sample. It is likely to contain drivers with less varied views (more homogeneous) than the population of all UK tanker drivers.

2 As Question **1**. Use where the population can be divided naturally into clusters and travelling or other difficulties make random or stratified sampling impractical.

Cluster sampling eliminates the possibility of bias by using random selection. Unlike quota sampling it does not ensure that different strata in the population are fairly represented in the sample.

3 **(a)** Cluster;

(b) No, all branches would be equally likely to be selected and so members in the smaller branches would have more chance of being included;

(c) All the branches had an equal number of members.

EXERCISE 2E

1 The sample is taken at regular intervals, for example a local radio reporter interviews every 200th person in a queue waiting to buy tickets for a football match.

2 Number employees 0000 to 2699. Choose a starting point between 0000 and 0053. Select this employee and every 54th thereafter. For example, if 0023 is chosen as the starting point choose employees numbered 0023, 0077, 0131, ... , 2669.

3 **(a)** Systematic;

(b) Yes, all have a probability of $\frac{1}{5}$ of being selected;

(c) Not all subsets of size 40 could be chosen. For example two friends entering the library together could not both be included;

(d) Likely to provide a useful estimate. Only unsatisfactory if the average number of books borrowed by customers who come early in the day is different from the average for the population as a whole or if the particular morning chosen is for some reason untypical.

EXERCISE 2F

1 **(a)** Number part time students 00 to 11, select two-digit random numbers; ignore repeats and >11; continue until four numbers chosen; select corresponding students. Choose eight of the full-time students in the same way.

(b) $\bar{x}$, there is a clearly a difference between the average age of part-time and full-time students. The stratified sample will ensure that the two strata are fairly represented in the sample. This might or might not be the case for the random sample.

2 **(a)** Yes;

(b) No, not all subsets of size 50 could be chosen;

(c) Likely to be representative. Only unrepresentative if customers who come early in the day have different eating habits, as a group, from all customers or if, say, weekend customers had different eating habits from weekday customers.

3 (a) (i) A random, B systematic, C stratified,
(ii) A equally likely, B equally likely, C equally likely;

(b) No, not all subsets of size 128 of the population are possible;

(c) In a quota sample the names from each of the strata are selected by any convenient method. This is much easier and quicker, in practice, than selecting a random sample from each of the strata;

(d) The stratification is clearly relevant to the question in that the electors from the different strata are likely to have different views on local authority housing. Stratified sampling ensures that all strata are fairly represented in the sample and so is to be preferred to random sampling which does not;

(e) It is not obvious that the different strata will have different attitudes to the monarchy (although they may do). If there is no difference then extra work involved in stratified sampling is to no purpose.

4 (a) (i) Yes, each house has probability of $\frac{1}{4}$,
(ii) No, not all subsets of the population possible;

(b) Not all members of a household equally likely to answer the door, not all subsets of electors possible or chosen houses are not a random sample;

(c) Number residents 00 to 62; select two-digit random numbers; ignore repeats and >62; continue until seven numbers obtained; choose corresponding residents.

5 (a) C;

(b) A cluster, B stratified, D random;

(c) A no, B yes, C no (or depends on method of choosing sample), D yes;

(d) (i) B ensures branches fairly represented in sample and avoids possibility of bias,
(ii) C easier to carry out.

6 (a) • Number names of first class passengers 00 to 69
• Select 14 two-digit random numbers (ignoring numbers >69 and repeated numbers)
• Choose first class passengers corresponding to the random numbers
• Number names of standard class passengers 000 to 179
• Select 36 three-digit random numbers (ignoring numbers >179 and repeated numbers)
• Choose standard class passengers corresponding to the random numbers

(b) (i) Suggestion B random. For suggestion A, not all samples of 45 possible, e.g. passengers in sample could not all be sitting in seats with different numbers.

(ii) All passengers have a probability of $\frac{45}{216} = \frac{5}{24}$ of being included in the sample for suggestion A.

7 (a) (i) Method 1 – stratified   Method 2 – cluster,

(ii) Method 1 – all teachers have a probability of $\frac{1}{20}$ of being included.
Method 2 – teachers in schools with a small number of teachers have more chance of being included than those in schools with a large number of teachers;

(b) (i) Not all samples of size 90 are possible, e.g. all males is not possible.

(ii) Males and females as groups are likely to have different views on maternity leave. This method ensures fair representation of each group.

(iii) When there is no difference in views on maternity leave between males and females.

(c) Advantage – easier and cheaper to carry out. Disadvantage – the teachers come from only 10 schools, so their views are likely to be more homogenous than those of all teachers in the authority.

3 Discrete probability distributions

EXERCISE 3A

1 1, 1, 1.

2 (a) 2.3, 8.41, 2.9;   (b) 113.3.

3 (a) 0.1;   (b) 1.5, 1.45, 1.20;   (c) 37.3.

4 (a) 2, 1.41.

5 (a) 0, 1, 1;   (b) 1, 1, 1, 2.

6 (a) 1.7, 1.18;   (b) 4.76.

7 (a) 4, 2.83;   (b)

x	0	5
$P(X = x)$	0.5	0.5

8 (a) 68.8, 21.6;   (b) 0.0464.

9 (a) 1.34, 1.83;   (b) (i) binomial,   (ii) 1.25, 0.968;

(c) Means fairly similar but standard deviations very different. Suggests distribution not binomial – applicants not guessing.

10 (a) 3.62, 1.95;   (b) $1.82 + 3.5p$, increase in sales if $p > 0.514$.

11 (a) 0.55;
(b) (i) 1.84,   (ii) 4.9,   (iii) 1.23;
(c) 0.0967.

12 (a) (i) 1.745,   (ii) 3.5825,   (iii) 0.733;
(b) (i) 0.44,   (ii) 1;
(c) (i) decrease – charge for children under 5 (zero) is less than for any other status,
(ii) increase – distribution more spread.

4 Poisson distribution

Answers have been given to four decimal places. However three significant figures is sufficient.

EXERCISE 4A

1 (a) 0.8472;   (b) 0.8488;   (c) 0.4457;
(d) 0.5928;   (e) 0.1377.

2 (a) 0.2381;   (b) 0.8893;   (c) 0.1954;
(d) 0.7108;   (e) 0.3712.

3 **(a)** 0.1048;   **(b)** 0.0895;   **(c)** 0.1013;
     (d) 0.6894;   **(e)** 0.7720.

4 **(a)** 0.0403;   **(b)** 0.7479;   **(c)** 0.7313;
     (d) 0.1067;   **(e)** 0.2480.

5 **(a)** 0.4628;   **(b)** 0.1607.

6 **(a)** 0.6728;   **(b)** 0.1463.

7 **(a)** 0.8335;   **(b)** 0.0394;   **(c)** 0.4082.

EXERCISE 4B

1 3.

2 **(a)** 13;   **(b)** 15;   **(c)** 17.

3 **(a)** 17;   **(b)** 21.

4 **(a)** 20;   **(b)** 23;   **(c)** 24.

EXERCISE 4C

1 **(a)** 0.3374;   **(b)** 0.2560;   **(c)** 0.1377;
     (d) 0.5928.

2 **(a)** 0.1665;   **(b)** 0.0985;   **(c)** 0.6575;
     (d) 0.0479;   **(e)** 0.1757.

3 **(a)** 0.9921;   **(b)** 0.1429;   **(c)** 0.6512;
     (d) 0.2424.

4 **(a)** 0.9909;   **(b)** 0.5940;   **(c)** 0.6919;
     (d) 0.2224;   **(e)** 0.0651.

5 **(a)** 0.2090;   **(b)** 0.1378;   **(c)** 0.3712;
     (d) 0.7668;   **(e)** 0.8444;   **(f)** 0.4075.

EXERCISE 4D

(a) Yes, Poisson likely to be valid;

(b) Where there is congestion lorries won't 'flow' independently so Poisson won't be valid;

(c) Unlikely as cars won't pass the point independently;

(d) Not Poisson, since there is an upper limit on the number of components;

(e) Yes, Poisson likely to be valid;

(f) Likely to be Poisson (although near Christmas the average rate is likely to change);

(g) Poisson likely;

(h) Not Poisson, mean not constant;

(i) Poisson likely;

(j) Poisson likely;

(k) If more than one person injured in an accident, injuries not independent, probably not Poisson.

MIXED EXERCISE

1 **(a)** 0.1653;   **(b)** 0.2694.

2 **(a)** 0.7787;   **(b)** 0.1254.

3 0.2213.

4 (a) (i) 0.5768, **(ii)** 0.9881, **(iii)** 0.0116;

   (b) 18.

5 (a) 0.2149; **(b)** 0.0424;

   Rate 1.6 per hour.

6 (a) The rate of arrival will vary – probably higher near 'mealtimes', lower at night. Not Poisson;

   (b) Not Poisson, calls can't arrive independently and at random;

   (c) Not Poisson since there is an upper limit of 150.

7 (a) (i) 0.3027, **(ii)** 0.6665, **(iii)** 0.0273;

   (b) 0.0290;

   (c) Demand may vary its rate with the seasons or for different days of the week.

8 (a) 0.5305;

   (b) The average rate will vary e.g. more on birthdays or near Christmas.

9 (a) (i) 0.7306, **(ii)** 5;

   (b) (i) Poisson. Conditions fulfilled.

      (ii) Not Poisson. Calls interfere with each other. Not random.

      (iii) Not Poisson. Mean not constant.

10 (a) (i) 0.8335; **(ii)** 0.3081;

   (b) 0.7787;

   (c) 0.0909;

   (d) Mean may vary according to day of the week.

11 (a) (i) 0.3012; **(ii)** 0.2169;

   (b) 0.0160;

   (c) 0.8666;

   (d) Mean may vary according to time of day. Also some cars will have more than one passenger → not independent.

5 Interpretation of data

EXERCISE 5A

1

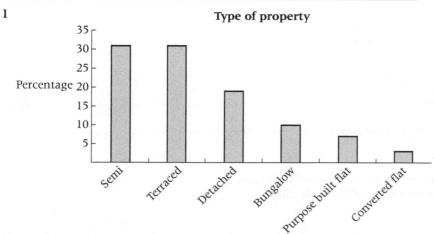

2

Drink	Frequency	Angle
Coffee	8	125
Orange	5	78
Tea	10	157
Total	23	

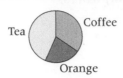

Pie chart to show types of drinks sold from 11:00 to 12:00 hours

Drink	Frequency	Angle
Coffee	24	122
Orange	38	193
Tea	9	46
Total	71	

Drink sales from 12:00 to 13:00 hours

Note: radius for 11.00–12.00 is 0.7 cm

so radius for 12.00–13.00 is $\sqrt{\dfrac{71}{23}} \times 0.7 = 1.2$ cm

3 (a)

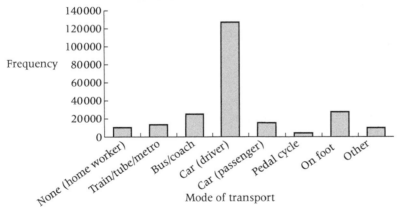

Bar chart to show mode of travel to work

(b) **Pie chart to show mode of transport**

Pie chart illustrates the **proportion** of people in each category.
Bar chart illustrates the **number** of people in each category.

EXERCISE 5B

1

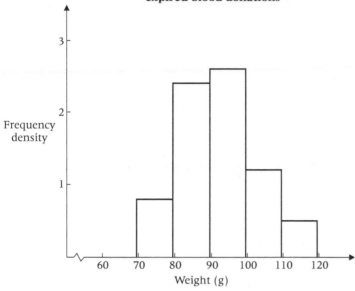

Histogram to show the weight of a sample of expired blood donations

Weight (g)	70–79	80–89	90–99	100–109	110–119
Frequency	8	24	26	12	5
Interval width	10	10	10	10	10
Frequency density	0.8	2.4	2.6	1.2	0.5

As all classes are of equal width the shape would be the same if frequency had been plotted on the vertical axis.

2

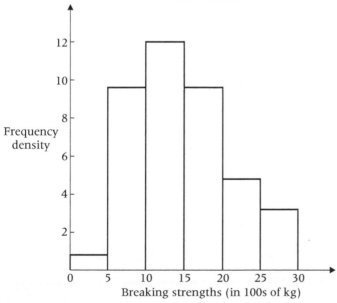

Histogram to show the breaking strain of 200 cables

Breaking strength	0–	5–	10–	15–	20–	25–30
Frequency	4	48	60	48	24	16
Interval width	5	5	5	5	5	5
Frequency density	0.8	9.6	12	9.6	4.8	3.2

All classes of equal width as in question 1.

3

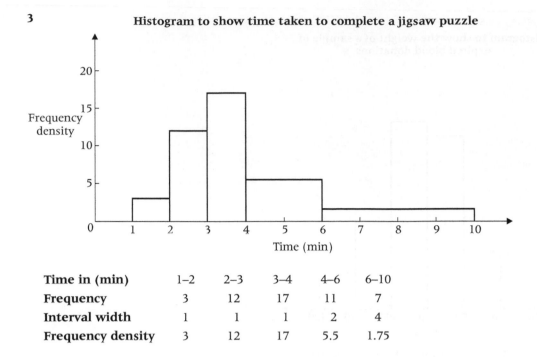

Histogram to show time taken to complete a jigsaw puzzle

Time in (min)	1–2	2–3	3–4	4–6	6–10
Frequency	3	12	17	11	7
Interval width	1	1	1	2	4
Frequency density	3	12	17	5.5	1.75

4

Age	0–4	5–15	16–24	25–44	45–74	75 and over
Frequency	4462	12 214	10 898	19 309	22 820	3364
Interval width	5	11	9	20	30	25*
Frequency density	892.4	1110.4	1210.9	965.45	760.6	134.56

*Assuming an upper age limit of 100 years.

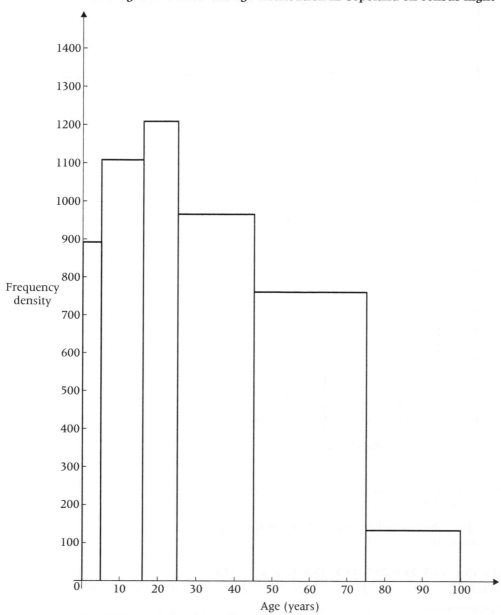

Histogram to show the age distribution in Copeland on census night

5 (a) A pie chart;

(b)

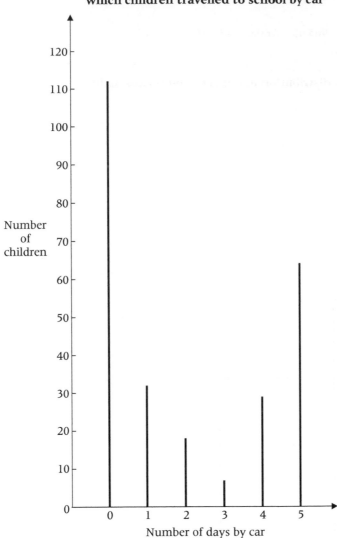

Line graph to show the number of days on which children travelled to school by car

Number of children *(y-axis)*

Number of days by car *(x-axis)*

Note: in order to see the line for 0 days it has been moved across from where the origin usually is.

Distribution is U-shaped, shows children tend to travel by car either every day or not at all.

(c)

Time (min)	0.5–15.5	15.5–25.5	25.5–35.5	35.5–55.5	55.5–90.5
Frequency	129	52	34	26	21
Interval width	15	10	10	20	35
Frequency density	8.6	5.2	3.4	1.3	0.6

Histogram to show the journey times

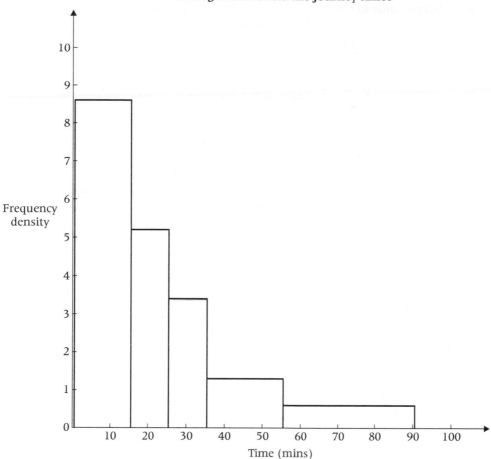

Time (mins)

EXERCISE 5C

1 When in order the data read

| 48 | 49 | 59 | 59 | 61 | 63 | 66 | 67 | 70 | 72 |

| 74 | 74 | 77 | 81 | 82 | 86 | 102 | 165 | 229 |

(a) the median is 72;

(b) the lower quartile is 61, the upper quartile is 82;

(c) boundaries for outliers are $61 - 1.5 \times (82 - 61) = 29.5$ and
$82 + 1.5 \times (82 - 61) = 113.5$, so 165 and 229 are outliers;

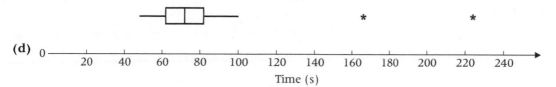

(d)

Time (s)

2 When sorted the data read

| 57.7 | 58.1 | 58.2 | 58.4 | 58.7 | 58.8 | 58.9 | 59.3 |

| 59.4 | 59.4 | 59.8 | 60.1 | 60.3 | 60.4 | 61.0 |

(a) For Brand A, smallest value 57.7, lower quartile 58.4, median 59.3, upper quartile 60.1, largest value 61.0.

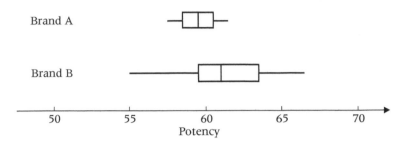

(b) The median potency for Brand B is greater than for Brand A, i.e. on average Brand B is more potent.

The variability of the potency for Brand A is smaller than that for Brand B.

3 (a) Median waiting time is the same for both offices.

B has three large outliers.

Ignoring these B has lower mean and less variability.

A is positively skewed.

B is negatively skewed.

(b) (i) The outliers were due to a cause outside post office's control – choose B for quicker service.

(ii) Outliers post office's responsibility, may recur. B generally quicker but there may be some very long waits. A avoids the very long waits.

4 (a) C's range (variability) is greater than that of D.

C's median time is greater than that of D.

C's times have positive skew, D's times have negative skew.

D will always beat C in a race on the basis of times in the box and whisker plot.

(b) (i) Choose A, since C's times are almost always longer than A's.

(ii) Choose B, since B is more likely to produce a very short time than A. A will hardly ever have a time overlapping the times taken by D.

(c) A.

EXERCISE 5D

1

Upper boundary	19.5	22.5	25.5	28.5	31.5	34.5
Cumulative frequency	0	3	9	21	30	32

(a)

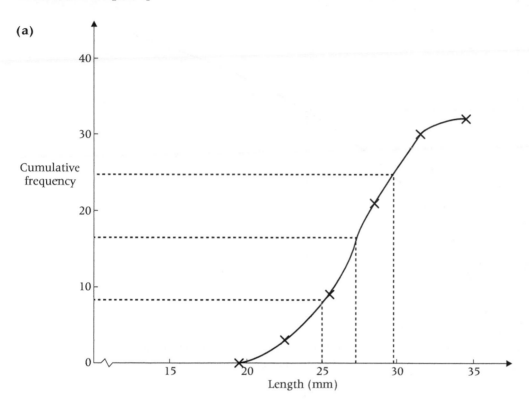

(b) median $= \dfrac{32 + 1}{2}$th = 16.5th reading = 27.2 mm

lower quartile $= \dfrac{32 + 1}{4}$th = 8.25th = 25.0 mm

upper quartile $= \dfrac{3(32 + 1)}{4}$th = 24.75th = 29.8 mm

Answers estimated from graph, there may be disagreement about the third significant figure.

2	**Upper boundary**	399.5	900.5	1500.5	3500.5	8000.5	20 000.5
	Cumulative frequency	0	5	14	54	99	159

$$\text{median} = \frac{159 + 1}{2}\text{th} = 80\text{th} = 6000 \text{ ohms/cm}$$

$$\text{lower quartile} = \frac{159 + 1}{4}\text{th} = 40\text{th} = 2800 \text{ ohms/cm}$$

$$\text{upper quartile} = \frac{3(159 + 1)}{4}\text{th} = 120\text{th} = 11\,100 \text{ ohms/cm}$$

3

Upper boundary	0	400	800	1200	1600	2000	3000	4000	5000	6000	8000
Cumulative frequency	0	25	56	100	157	231	389	444	470	488	500

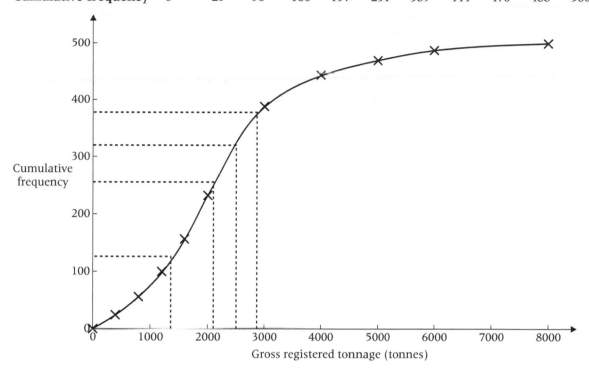

(a) median $= \dfrac{500 + 1}{2}$th $= 250.5$th $= 2100$ tonnes

(b) lower quartile $= 125.25$th $= 1380$ tonnes

upper quartile $= 375.75$th $= 2880$ tonnes

∴ interquartile range $= 2880 - 1380$
$= 1500$ tonnes

(c) % exceeding 2500 tonnes $= \dfrac{(500 - 320)}{500} \times 100 = 36\%$.

EXERCISE 5E

1 (a)

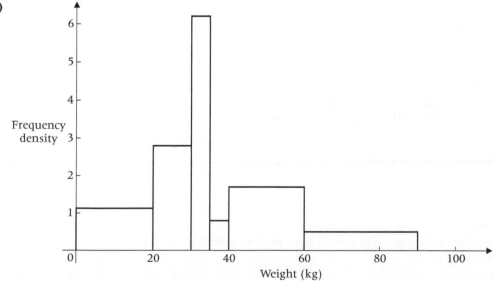

(b) Bimodal distribution, probably because passengers who would have taken just over 35 kg of luggage have taken less to avoid excess charge.

(c)

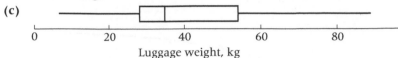

Luggage weight, kg

(d) The box-and-whisker plot does not illustrate the feature described in **(b)**. The histogram is a better diagram for this data.

2 (a)

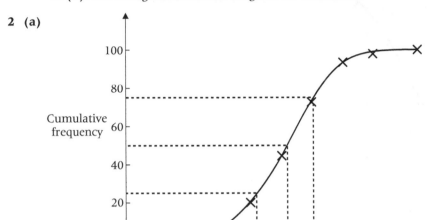

(b) Estimated median 131 seconds, interquartile range 38 seconds;

(c) Second jigsaw takes longer on average so more difficult, also times taken are more variable;

(d) 137 seconds.

It is possible to estimate the median of the times of the 116 children. It is not possible to calculate the arithmetic mean as the times for the slowest 16 children are unknown.

3 (a) A has lowest variability, median just over 5, apart from 4 outliers B has low variability and lowest median, C has large variability and median close to 6, it is negatively skewed.

(b) Choose A as it has low variability and no outliers. The mean can be adjusted to bring it to the required value.

4 (a) 1280;

(b) (i) bar chart,

(ii) Very small proportion of Victoria Records' sales. Similar proportion of Ron's sales to tapes and records.

5 (a) 80, 1.4;

(b) (i) Employment status is qualitative not quantitative,

(ii) 239°,

(iii) 1.3 cm.

6 (a) 105.9 million;

(b) (i) The number of sites seems to exhibit some random variation but no trend,

    (ii) The number of screens shows an upward, approximately linear trend.
The average number of screens per site has increased steadily from just over two in 1988 to just over four in 1998.

(c) £2.77;

(d) The revenue per admission has increased from £1.89 to £3.64 from 1988 to 1998, an increase of 93%.
The retail price index has increased from 106.6 to 163.4, an increase of 53%. There has been a substantial increase in admission prices even allowing for the effect of inflation.

7 (a) 9.7 thousand tonnes;

  (b) £7002 thousand;

  (c) (i) Quantity of plaice landed has shown a downward trend with some indication that this trend has now levelled out,

    (ii) Quantity of sand eels landed has shown a steep but erratic upward trend;

  (d) (i) £743 per tonne,   **(ii)** £542 per tonne;

  (e) (i) £3260 per tonne,   **(ii)** £4490 per tonne;

  (f) Since the quantity of brill caught is given to only 1 sf the actual quantity is between 0.45 and 0.55. There could be an error due to rounding in the answers to **(e)** of up to about 10%.

  (g) There was an increase in the RPI of 13% between 1992 and 1997. There has been a reduction of 27% in the average price of haddock. In real terms this reduction is even greater. There has been an increase of about 38% in the average price of brill. This comfortably exceeds the increase in the RPI (even allowing for possible rounding errors) and so represents a real increase in the price.

8 (a) The number of sites shows an upward trend. The rate of increase has been accelerating up to 1993. From 1993 to 1996 the rate of increase has reduced;

  (b) The number of councils participating showed an upward, approximately linear trend until 1990 but then levelled off;

  (c) Local government reorganisation reduced the number of councils in 1996;

  (d) In 1979 only a small proportion of councils were participating whereas in 1985 a large proportion of councils were participating. It is therefore impossible to maintain the rate of increase in the number of councils participating;

  (e) 1977 – 2.8, 1982 – 5.3, 1988 – 8.8, 1995 – 31.2.
As well as an increase in the number of councils participating there has also been a marked increase in the number of sites per participating council.

9 (a) 30%;

  (b) 22%;

  (c) The bigger the household income the bigger the proportion of women who have never smoked;

  (d) As with women the bigger the household income the bigger the proportion of men who have never smoked. For any given household income group, the proportion of men who have never smoked is smaller than the proportion of women who have never smoked;

10 (a) (i) 4.1 million tonnes;
  (b) (i) Downward, approximately linear trend,
    (ii) Relatively little use of gas up to 1992 then a marked upward trend,
    (iii) A relatively slow upward trend.
  (c) 6.3°. This sector would have been much larger in 1988 (32.2°);
  (d) Since the figure for hydro is given to only 1 sf the use in each of the years 1988 to 1991 may not have been the same but could have varied between 0.35 and 0.45 (million tonnes of oil equivalent). The increase from 0.4 to 0.5 might be almost entirely accounted for by rounding error;
  (e) 1.9 million tonnes.

11 (a) (i) 4.5 g, **(ii)** 95.5 g, **(iii)** 91 g;
  (b) (i) 0.0401, **(ii)** 0.000 03;
  (c) The average contents are less than the nominal quantity, more than 2.5% are non-standard. Requirements not met.
  (d) More than 2.5% non-standard (8.5%), more than 0.1% inadequate (0.6%). Conditions not met.

12 (a) 18 536 000;
  (b) £478;
  (c) Advantage – it eliminates affect of inflation from the data. Disadvantage – the amounts shown for years other than 1995 are not the actual expenditure;
  (d) The ratio of constant 1995 prices to current prices should be the same in any given year. In the table, this ratio is <1 for spending in the UK in 1998, but >1 for spending overseas;
  (e) • The number of visits has increased from 1988 to 1998 (both to and from the UK).
    • Expenditure has increased from 1988 to 1998 both in actual and in real terms.
    • More UK residents visit overseas than overseas residents visit the UK.
    • Average expenditure of overseas residents in the UK has increased in real terms.

13 (a) 7 367 000;
  (b) 563;
  (c) Expenditure from public funds has increased by about 26% from 1993–94 to 1998–99. The Retail Price Index has increased by less (16%), showing a real increase in expenditure from public funds.
  (d) • The total number of dentists has steadily increased from 1993–94 to 1998–99.
    • The number of registered patients has decreased (both adults and children); particularly in the last year.
    • The number of courses of treatment has increased.
    • The increase in expenditure is similar to the increase in the RPI, i.e. there is little change in real terms.

6 Hypothesis testing

EXERCISE 6A

1 ts 1.80   cv 1.6449   significant evidence that the mean is greater than 135 kg.

2 ts −1.22   cv ±1.96   no significant evidence to doubt that the pop mean reaction time is 7.5 s.

3 ts 1.05   cv 1.6449   no significant evidence that children take longer.

4 ts −3.48   cv ±2.5758   significant evidence that the pop mean length is not satisfactory (less than 2 cm).

5 ts 1.67   cv 1.6449   significant evidence that the silver is not pure.

6 ts 3.19   cv 2.3263   significant evidence that the pop mean weight has increased.

7 ts −1.57   cv −1.6449   no significant evidence that pop mean weight has reduced.

8 ts 3.95   cv ±1.96   significant evidence that the pop mean time is greater than 4 s.

EXERCISE 6B

1 ts 2.42   cv 2.3263   significant evidence that the pop mean breaking strength is greater than 195 kg.

2 ts 1.02   cv ±1.96   no significant evidence to doubt that the pop mean resistance is 1.5 ohms.

3 ts −2.67   cv −1.6449   significant evidence to suggest that the pop mean test score is lower.

4 ts 1.79   cv 1.6449   significant evidence to suggest that the pop mean weight is greater than 500 g.

5 ts −2.19   cv −2.3263   no significant evidence that the customer's suspicions are correct.

Sample assumed random.

6 ts 1.87   cv 2.3263   no significant evidence that the pop mean weight is greater than 35 g.

7 (a) $H_0 \mu = 7.4$ $H_1 \mu < 7.4$ ts −4.98   cv −1.6449 significant evidence that the mean is less than 7.4;

(b) Test statistic is −4.98 so conclusion would have been unchanged even if the significance level had been very much less than 5%. Conclusion is very clear;

(c) As the sample is large it is not necessary to make any assumption about the distribution. However it is necessary to assume the sample is random. There is no information on this. If, for example, only those with particular symptoms had been tested they might be untypical of all varicoceles sufferers.

EXERCISE 6C

1 (a) (i) Conclude pop mean greater than 135 kg when in fact pop mean equals 135 kg,

  (ii) Conclude pop mean equals 135 kg when in fact pop mean greater than 135 kg;

  (b) (i) 0.05,   **(ii)** 0.

2 (a) (i) Conclude mean unsatisfactory when it is satisfactory,

  (ii) Conclude mean is satisfactory when it is unsatisfactory;

  (b) (i) 0.01,   **(ii)** 0.

3 (a) ts -1.80 cv ± 1.96 accept pop mean time is 20 minutes;

  (b) Conclude pop mean time not 20 minutes when in fact it is 20 minutes;

  (c) Would need to know the actual value of the population mean.

4 (a) ts -1.69 cv ± 1.6449 significant evidence pop mean length not (less than) 19.25 mm;

  (b) (i) Conclude pop mean not 19.25 mm when in fact it is 19.25 mm,

  (ii) Conclude pop mean 19.25 mm when in fact it is not 19.25 mm;

  (c) 0.1.

5 (a) ts -1.90 cv -1.6449 significant evidence population mean less than 47, easier to assemble;

  (b) Conclude not easier to assemble when in fact it is.

7 Contingency tables

Values of X^2 may vary slightly according to the number of decimal places used for E.

EXERCISE 7A

1 $X^2 = 38.7$ cv 9.21   time of day associated with type of birth.

2 $X^2 = 23.8$ cv 9.21   proportion favouring road improvements not independent of area.

3 $X^2 = 6.67$ cv 7.815   accept choice of dish independent of day of week.

4 (a) Only women in the main shopping area asked (many other answers possible);

  (b) $X^2 = 17.7$ cv 9.49   respondents' replies not independent of age.

5 (a) $X^2 = 17.0$ cv 5.99   proportion transferred not independent of surgeon;

  (b) B had substantially less transferred than expected. Need to know how cases were allocated. It could be, say, that only straightforward operations were undertaken by B.

6 (a) $X^2 = 11.4$ cv 9.21   sex and grade not independent;

  (b) Females did better – more than expected obtained high grades and less than expected had low grades;

  (c) (i) Need total number of candidates in order to calculate frequencies,

  (ii) cv 10.6   subject and grade not independent,

(iii) A bigger proportion of statistics candidates than pure and applied candidates obtained high grades. Would need some information on quality and preparation of candidates to say whether it is easier to get a good grade in statistics.

7 $X^2 = 16.9$ cv 13.3

Grade and length of employment not independent. Unskilled staff tend to stay longer than skilled staff. In particular more unskilled employees than expected stay longer than five years and more skilled staff than expected leave within two years.

EXERCISE 7B

1 $X^2 = 12.4$ cv 5.99   age and sex not independent.

2 $X^2 = 3.14$ cv 7.815   (5%) (another % could be chosen as not specified in question) accept choice independent of gender.

3 $X^2 = 1.99$ cv 9.21   accept method of communication independent of subject.

4 (a) $X^2 = 35.3$ cv 7.815   proportion of guests rating a feature important not independent of feature;

(b) Availability of squash courts much less important than other features. Little to choose between other features – number rating them important exceeded expected number by roughly the same proportion;

(c) If $E < 5$;

(d) Comfortable beds – most similar category;

(e) cv between 73.3 and 79.1. Rating and feature not independent.

EXERCISE 7C

1 $X^2 = 5.73$ cv 3.84   proportion exhibiting allergies not independent of treatment. Drug effective – less than expected of those receiving the drug exhibited allergies.

2 $X^2 = 3.50$ cv 3.84   (5%) (another % could be used) accept proportion germinating independent of variety.

3 $X^2 = 0.618$ cv 3.84   accept proportion defective independent of mould.

4 $X^2 = 1.06$ cv 3.84   accept equal proportion of librarians and designers can distinguish the word.

5 (a)

	<87%	≥87%
National	10	5
Labour	6	7

(b) $X^2 = 0.507$ cv 3.84   accept winning party independent of turnout;

(c) Constituencies appear to have been selected in alphabetical order. No constituencies from last part of alphabet. This could effect the test but election results unlikely to be associated with name of constituency so test probably valid. (Other answers possible.)

MIXED EXERCISE

1 (a) $X^2 = 8.11$ cv 3.84   reason associated with time of day;

(b) Use of the crossing because of The Beatles' association is more likely out of the rush hour.

2 (a) $X^2 = 24.5$ cv 11.3   classification not independent of inspector;

(b) Judgements are inconsistent. de Sade classified a smaller proportion as unsatisfactory than did the others. The results should be taken seriously;

(c) Test invalid – Os are not frequencies.

3 (a)

Week	1	2	3	4	5
Statistics tables	24	32	20	18	9
Other items	192	168	146	87	55

(b) $X^2 = 3.56$ cv 9.49   accept proportion of items which were statistical tables independent of week.

4 (a) $X^2 = 7.59$ cv 3.84   colour not independent of habitat;

(b) Greater proportion of woodland snails dark;

(c) Test valid since Es > 5;

(d) Test not valid since Os are not frequencies.

5 (a) $X^2 = 18.9$ cv 5.99   returning 1997 questionnaire not independent of answer to 1996 question on truancy;

(b) Less likely. Less than expected truants returned the questionnaire;

(c) Truants less likely to return questionnaire. Respondents may not tell the truth.

6 (a) $X^2 = 72.8$ cv 7.815   snoring associated with heart disease;

(b) Snore more e.g. more sufferers snore every night than expected;

(c) No. Heart disease and snoring associated but there is no proof of cause and effect.

7 (a) $X^2 = 33.5$ cv 7.815   outcome not independent of treatment;

(b) New treatment effective since number of those receiving new treatment who show marked improvement greater than expected;

(c)

	Died	Refused	Untraceable
New	19	10	15
Standard	3	12	18

cv 5.99   reason depends on treatment.

(d) New treatment has a high risk of death and so should not be used. This was hidden in first set of data where patients who died were included in the category 'information unobtainable'.

8 Distribution-free methods for single samples and paired comparisons

EXERCISE 8A

1 H_0 pop med $= 135$, H_1 pop med > 135.
One-tail 5%, $n = 14$, ts $= 10^+$ or 4^-.
$P(x \leqslant 4^-) = P(x \geqslant 10^+) = 0.0898 > 0.05$ (one-tail).
Accept H_0. No significant evidence to doubt pop med breaking strength is 135.

2 H_0 pop med $= 7.5$, H_1 pop med $\neq 7.5$.
Two-tail 5%, $n = 25$, ts $= 7^+$ or 18^-.
$P(x \leqslant 7^+) = P(x \geqslant 18^-) = 0.0216 < 0.025$ (two-tail).
Reject H_0. Significant evidence pop med reaction time not 7.5 (lower).

3 H_0 pop med $= 15$, H_1 pop med < 15.
One-tail 1%, $n = 20$, ts $= 3^+$ or 17^-.
$P(x \leqslant 3^+) = P(x \geqslant 17^-) = 0.00129 < 0.01$ (one-tail).
Reject H_0. Significant evidence that pop med number remembered is less than 15.

4 H_0 pop med $= 7.4$, H_1 pop med > 7.4.
One-tail 5%, $n = 9$, ts $= 7^+$ or 2^-.
$P(x \geqslant 7^+) = P(x \leqslant 2^-) = 0.0898 > 0.05$ (one-tail).
Accept H_0. No significant evidence to doubt pop med time to complete for children is 7.4.

5 H_0 No difference in preference for sunflower and olive oil.
H_1 Sunflower preferred.
One-tail 5%, $n = 30$, ts $= 20^+$ or 10^-.
$P(x \geqslant 20^+) = P((x \leqslant 10^-) = 0.494 < 0.05$ (one-tail).
Reject H_0. Significant evidence that sunflower is preferred.

6 (a) H_0 population median $= 52$, H_1 population median > 52.
One-tail 1%, ts $= 18^+$ or 3^-.
$P(\geqslant 18^+) = P(\leqslant 3^-) = 0.000\,745 < 0.01$.
Reject H_0. Significant evidence that new recipe is preferred;

(b) Less than half preferred original flavour, accept original flavour **not** preferred without further calculation.

7 H_0 No difference H_1 A preferred   One-tail 5%.

(a) ts $= 21^+$ or 9^- ignore 2 zero $n = 30$.
$P(x \geqslant 21^+) = P(x \leqslant 9^-) = 0.0214 < 0.05$ (one-tail).
Reject H_0. Significant A better;

(b) Features randomly selected. Features equally important. These are only relevant features to select car quality.

8 (a) H_0 population median $= 12$, H_1 population median > 12.
$n = 30$, $P(23$ or more$) = 0.0026$. Reject H_0, at 1% significance level, conclude average (as measured by median) annual number of visits to the cinema exceeds 12;

(b) (i) Conclusion unaffected, the sign test does not assume normal distribution,

(ii) Conclusion unaffected, the sign test does not assume symmetrical distribution,

(iii) Data biased as all interviewees had been to the cinema at least once. Conclusions completely unreliable.

EXERCISE 8B

1 $T^- = 27\frac{1}{2}$, $T^+ = 77\frac{1}{2}$, H_0 Med/mean pop $= 135$, $n = 14$,
ts $= 27\frac{1}{2}$,   cv $= 26$, H_1 Med/mean pop > 135,   one-tail 5%.
Accept H_0. No significant evidence to doubt med $= 135$.

2 $T^- = 9.5$, $T^+ = 35.5$, H_0 Med/mean pop $= 7.4$, $n = 9$,
ts $= 9.5$, cv $= 8$, H_1 Med/mean pop > 7.4, one-tail 5%.
Accept H_0. No significant evidence to doubt med $= 7.4$.

3 $T^- = 75$, $T^+ = 3$, H_0 Med/mean pop $= 51$, $n = 12$,
ts $= 3$,   cv $= 10$, H_1 Med/mean pop < 51, one-tail 1%.
Reject H_0. Significant evidence that med/mean birth rate is less than 51.

4 $T^- = 20$, $T^+ = 35$, $\mathbf{H_0}$ Med/mean pop $= 23$, $n = 10$,
    ts $= 20$,   cv $= 8$, $\mathbf{H_1}$ Med/mean pop $\neq 23$,   two-tail 5%.
    Keep $\mathbf{H_0}$. No significant evidence to doubt med score in test is 23.

5 $T^- = 91$, $T^+ = 45$, $\mathbf{H_0}$ pop med/mean $= \$95$, $n = 16$,
    ts $= 45$,   cv $= 30$, $\mathbf{H_1}$ pop med/mean $\neq \$95$,   two-tail 5%.
    Accept $\mathbf{H_0}$. No significant evidence to doubt med/mean amount spent is $95.

6 **(a)** Population not symmetrical (batsmen cannot score less than 0 but can score a great deal more than 30).

    (b) $\mathbf{H_0}$ pop med $= 30$, $\mathbf{H_1}$ pop med < 30.
       $n = 20$, one-tail 5%, ts $= 5^+$ or 15^-,
       $P(\leq 5^+) = P(\leq 15^-) = 0.0207 < 0.05$ (one-tail).
       Reject $\mathbf{H_0}$. Significant evidence med score is less than 30.

7 **(a)** $\mathbf{H_0}$ pop med $= 1.5$, $\mathbf{H_1}$ pop med > 1.5.
       one-tail 5%, $n = 15$, ts $= 3^-$ or 12^+
       $P(\leq 3^-) = P(\geq 12^+)$
             $= 0.0176 < 0.05$ (one-tail).

       Reject $\mathbf{H_0}$. Significant evidence pop med relative risk is greater than 1.5;

    (b) $\mathbf{H_0}$ pop mean $= 74$, $\mathbf{H_1}$ pop mean > 74
       one-tail 1%

       $T^- = 1$, $T^+ = 35$, $n = 8$,
       cv $= 2$,   ts $= 1$,   ts $<$ cv

       Reject $\mathbf{H_0}$. Significant evidence male teachers have mean blood pressure above 74.

    (c) Only male teachers.
       Only teachers working in large comprehensive.

8 **(a)** $\mathbf{H_0}$ population median $= 200$, $\mathbf{H_1}$ population median > 200.
       Sign test $n = 20$, $P(14$ or more$) = 0.0577$, accept $\mathbf{H_0}$, no significant evidence that median life exceeds 200 minutes;

    (b) $\mathbf{H_0}$ population median $= 200$, $\mathbf{H_1}$ population median > 200.

Difference	-38	-21	-17	-16	-11	-5	10	15	25	34
	39	>40	>40	>40	>40	>40	>40	>40	>40	>40
Rank	-10	-7	-6	-5	-3	-1	2	4	8	9
	11	12	13	14	15	16	17	18	19	20

       ts $= 32$, cv for 1%, one-tailed test $= 43$.
       Reject $\mathbf{H_0}$ conclude median life greater than 200 minutes;

    (c) Sample random, distribution symmetrical;

    (d) Would only know that 13 positive differences exceeded 12. Would not have been able to allocate signed ranks as four of the negative differences are numerically greater than 12.

9 **(a)** $\mathbf{H_0}$ population median $= 20$, $\mathbf{H_1}$ population median $\neq 20$.
       ts $= 13$, cv for 5%, two-tailed test $= 8$, accept $\mathbf{H_0}$ that mean external diameter is 20;

    (b) $\mathbf{H_0}$ population median $= 20$, $\mathbf{H_1}$ population median $\neq 20$.
       ts $= 11$, cv for 5%, two-tailed test $= 8$, accept $\mathbf{H_0}$ that mean external diameter is 20;

    (c) **(i)** null hypothesis in this case states the value of the median which will be accepted unless there is overwhelming evidence to disprove it,
         (ii) Type I error is concluding that the median is not 20 when in fact it is 20. (A Type I error cannot have been made in this example as the null hypothesis has been accepted in both cases.)

(iii) Type II error is accepting that the median is 20 when in fact it is not 20. (At least one Type II error has been made in this example as the original measurements and the corrected measurements cannot both have a median of 20.)
Accepting the null hypothesis is not the same as proving it is true. The technicians conclusions merely state that in neither case is there overwhelming evidence to reject the null hypothesis.

EXERCISE 8C

1 P(2 or fewer +) + P(2 or fewer −) = 0.109, no significant evidence of difference.

2 **(a)** Each student has a pair of times – one with audience, one without;

(b) Differences between students are likely to be a major contributor to experimental error. This source is eliminated by using the differences in times for each student;

(c) P(2 or fewer +) = 0.0547 evidence of longer times in front of audience not (quite) significant;

(d) With audience always done after without audience. Order could effect result.
Population not defined, nor is method of selecting the sample.
Sample small – a larger sample might have lead to a significant difference.
Define population. Attempt to ensure unbiased sample (ideally random but this is unlikely to be possible).
Randomise order of tasks.

3 P(1 or fewer +) = 0.0352 significant evidence that new preparation leads to greater improvement.

4 P(3 or fewer +) + P(3 or fewer −) = 0.146 no significant evidence of difference in load times.

5 P(2 or fewer +) = 0.0193 significant evidence that EPCSD is quicker. Need information on costs, availability, training requirements, attitude of workforce, etc. before making a decision.

MIXED EXERCISE

1 **(a)** Paired design reduces experimental error, allocation at random reduces possibility of any difference found being due to order or change in conditions;

(b) T = 4, cv = 8, significant evidence that nature of surface influences athletes' performance (faster on synthetic track).

2 T = 6.5, cv = 11, significant evidence that students take longer with audience.

3 T = 1.5, cv = 2, significant evidence of greater improvement using new preparation.

4 T = 18, cv = 14, no significant evidence of difference.

5 **(a)** Reduces experimental error by eliminating differences between subjects, particularly useful in this case as differences between subjects appear to be large;

(b) T = 15.5, cv = 4, no significant evidence of alcohol influencing blood clotting time;

6 T = 16.5, cv = 17, significant evidence of difference (less wear using leather A).

7 T = 8.5, cv = 10, significant evidence zinc concentration higher on river bed.

8 (a) (i) P(2 or fewer +) + P(2 or fewer −) = 0.109, no significant evidence of difference,

(ii) T = 7, cv = 8, significant evidence of difference (pulse rate higher at 11 am);

(b) Wilcoxon takes account of magnitude as well as sign of the differences and so is more likely to detect a difference if one exists. In this case Wilcoxon has found significant evidence of a difference but the sign test did not.

9 (a) H_0 population median difference = 0
H_1 population median difference > 0.

	Mon	Tue	Wed	Thu	Fri	Sat
New − Old	18	5	−6	−2	42	19

T = 4, cv = 2 for 5%, one-tailed test.
Accept H_0 conclude no difference between suppliers.

(b) Takings will be affected by factors other than the supplier. This is known as experimental error. The difference found in Jim's original 1-day trial may be due to experimental error rather than to a difference between suppliers. In the analysis of Yasmin's trial the differences between takings on the same day of the week are analysed thus removing the effect of 'day' – a large source of experimental error. The analysis is based on more than one difference (replicates) thus allowing the size of the remaining experimental error to be taken into account. To remove the possibility of unconscious bias entering the experiment the order in which the two suppliers were used should be chosen at random.

10 (a) (i) H_0 population median difference = 0
H_1 population median difference ≠ 0.
$n = 8$, $p = 0.5$, P(5 or more +) + P(5 or more −) = 0.7266.
Accept H_0 no difference in time to spoilage.

(ii) H_0 population median difference = 0
H_1 population median difference ≠ 0.
T = 6, $n = 7$, cv for 5%, two-tailed test = 2.
Accept H_0 no difference in time to spoilage.

(b) Both tests assume the sample may be regarded as random.
Valid to use all eight pairs for sign test. Magnitude of one difference unknown so cannot be included in Wilcoxon's test. This test is valid for the remaining seven differences provided it is reasonable to assume that the population is symmetrical.
However neither test shows a significant difference.

(c) Advantages of sign test – quicker and easier to collect data. Only necessary to observe which half of each pair spoils first – no need to assume population symmetrical.
Advantage of Wilcoxon's test – more likely to detect a difference if such a difference exists.

11 (a) H_0 pop med difference = 0
H_1 pop med difference ≠ 0.
two-tail 5%
$n = 11$, $T^- = 57\frac{1}{2}$, $T^+ = 8\frac{1}{2}$
cv = 11,   ts = $8\frac{1}{2}$,   ts < cv

Reject H_0. Significant evidence to suggest a difference in median times;

(b) Wilcoxon signed-rank test requires that the sample is taken from a population that is symmetrically distributed (population of differences);

(c) Differences between the two groups of people might influence the results, rather than the effect of the alcohol alone.

12 (a) H_0 population median difference = 0
H_1 population median difference > 0.
$n = 14$, $p = 0.5$, P(9 or more +) = 0.2120.
Accept H_0 no significant difference in coursework and examination marks.

(b) Very little variation in coursework marks for those students who handed it in. Coursework marks only distinguish between those who handed in and the others.

(c) H_0 population median difference = 0,
H_1 population median difference > 0.
$n = 11$, $p = 0.5$, P(9 or more) = 0.0327.
Reject H_0 coursework marks significantly higher than examination marks on average.

(d) Those students who hand in coursework appear to all get about the same mark. This mark is higher, on average, than their examination marks. This difference disappears if all students are included in the analysis.

13 (a) H_0 population median difference = 0
H_1 population median difference > 0.
$n = 11$, $p = 0.5$, P(8 or more +) = 0.1133.
Accept H_0, proposal 2 not significantly better than proposal 1.

(b) Members of staff who read the proposals could be regarded as a random sample;

(c) H_0 population median difference = 0
H_1 population median difference ≠ 0.
T = 2, $n = 9$, cv for 5%, two-tail test = 6.
Reject H_0 proposal 2 better than proposal 1;

(d) Differences symmetrically distributed;

(e) No difference for sign test but Wilcoxon's signed-rank test could not have been carried out.

9 Distribution-free methods for two or more unpaired samples

EXERCISE 9A

1 U = 35, cv = 27. Accept no significant differences in average times for girls and boys.
Random selection of subjects good. Similar numbers of boys and girls OK, but might have been better to arrange to have same number of each. Population is one age group at a particular school so conclusions only apply to this limited group.

2 U = 6.5, cv = 11. Significant evidence that premium batteries last longer on average.
There is no control over how the cars (batteries) are used. This may lead to substantial experimental error but no obvious reason why it should lead to bias.

3 U = 12.5, cv = 4. Accept no significant difference between average scores for boys and girls.

4 U = 14, cv = 13. Accept no significant difference between average scores on the two tests.
Experimental error is in this case the effect of factors other than different tests on the results. If there were no experimental error all the scores on a particular test would be the same.

5 Mann–Whitney, U = 15, cv = 15. Significant evidence plants grow faster, on average, on the west facing side.
Mann–Whitney can be used to compare two populations where the data is not paired.

6 U = 9, cv = 8. Accept no significant evidence of difference between oranges and lemons.

7 **(a)** Two independent samples, no natural pairing;

   (b) U = 4, cv = 5. Evidence of higher cost in Southville.

8 **(a)** Mann–Whitney, U = 19, cv = 8. Accept no significant difference in times;

   (b) If route is used regularly drivers are unlikely to lose their way, therefore reasonable to exclude this result as purpose is to compare times in normal use.

9 **(a)** U = 8, cv = 5. Accept no significant difference in miles driven on 4 gallons of petrol;

   (b) Good features – relatively cheap and easy to collect data.
   Bad features   – samples self selecting – may involve bias,
                  – little opportunity to standardise driving conditions or ensure accuracy of data,
                  – no control over sample sizes.

10 **(a)** All oranges of variety P measured by inspector A and all variety B measured by inspector B. If each inspector had measured some of each, any systematic differences in their measurements could not have been confused with a difference between varieties of orange.

   (b) U = 24, cv = 27. Evidence of difference in average weights (P heavier).

11 **(a)** Mann–Whitney, U = 13, cv = 7. Accept no significant difference between average days to puncture;

   (b) Good features – data easy and cheap to collect.
   Bad features   – samples self selecting – may involve bias,
                  – little opportunity to check accuracy of data,
                  – distance cycled would be a more relevant variable (although more difficult to obtain) than time,
                  – no control over sample sizes. A larger sample for B would have given more chance of detecting a difference if one exists.

   (c) Type 1 error – concluding a difference between makes of tyres exists when it does not (cannot have been made in this case).
   Type 2 error – accepting no difference between makes of tyre when one exists.

12 **(a)** Choose Wilcoxon's signed-rank test since both possible and Wilcoxon gives more chance of detecting a difference if one exists.
   T = 1, cv = 0.
   Evidence not significant at 1%. Accept mean time 210 seconds.

(b) U = 2, cv = 5.
Conclude there is evidence of a difference in population average shaving times for the two people (the assistant is faster).

(c) Since there is a difference in average shaving times for different people, same person should test more than one type of razor. Standard and new design could both be used by several people and analysed using a paired comparison. Method of shaving and timing could also be standardised to further reduce experimental error.

13 (a) U = 11, cv = 11. Significant evidence of a difference in population average creatinine levels measured by the two machines (Analyser A are lower on average);

(b) T = 1, cv = 2. Significant evidence of a difference in population average creatinine levels measured by the two machines (Analyser A are lower);

(c) Since in second trial the same samples of blood were analysed, the lower measurements by Analyser A can no longer be explained by non-random differences between blood samples. However still unable to say which analyser is more accurate.

(d) Analyser A, T = 1.5, cv = 4. Significant evidence of bias. Analyser B, T = 10, cv = 4. Accept mean is 90 – no significant evidence of bias.

(e) (i) Analyser A shown to be biased, suggests use B to avoid bias,

(ii) Since all readings are on same sample low variability is desirable. Analyser B appears to be much more variable. Recommend use Analyser A after recalibration to remove bias.

14 (a) U = 16, cv = 21. Conclude significant evidence that apples of variety 2 are on average heavier than variety 1;

(b) No effect, rankings would still be the same;

(c) No effect, the evidence that variety 2 was heavier would be even stronger.

EXERCISE 9B

In questions 2, 3 and 4 the sample sizes are less than five and so the critical values may be unreliable.

1 (a) H = 13.3   cv = 9.21
Significant evidence to doubt H_0 and conclude that at least two of the populations differ.

(b) It appears that there certainly is a significant difference in population median reading scores (scores higher for high motivation) between children taught under low and high motivation.

2 (a) H = 1.88   cv = 5.99
No significant evidence to suggest that the populations differ. Accept H_0.
No significant evidence of any difference in average lengths of life.

(b) There is no indication of the amount of usage of the calculators being uniform for the staff involved in the trial.

3 (a) Suitable test is Kruskal–Wallis

$H = 4.27 \qquad cv = 5.99$

No significant evidence to suggest that the populations differ. Accept H_0.

(b) There is no significant evidence that the average level of absenteeism differs for people taking the existing or new drugs, or the placebo.

4 $H = 13.0 \qquad cv = 7.815$

Significant evidence to doubt H_0, and conclude that at least two of the populations differ.

It appears that there certainly is a significant difference in population average times (times lower for Group D, four experiences) between mice which have had only one experience and mice who have had four experiences.

5 $H = 8.49 \qquad cv = 4.605$

Significant evidence to doubt H_0, and conclude that at least two of the populations differ. Oven 1 hotter than oven 3.

6 (a) $H = 16.1 \qquad cv = 12.6$

Significant evidence to doubt H_0, and conclude that at least two of the populations differ.

(b) Darts players and 'None' did least well (mean placing about 38 in each case), significantly worse than Rugby and Tennis players (mean placing about 17 in each case).

(c) Assume the samples could be regarded as random samples from each category.

10 Correlation

EXERCISE 10A

1 (a) 0.997;

(b) $H_0 \; \rho = 0$, $H_1 \; \rho \neq 0$, test statistic 0.997, $n = 10$, 2 tail 5%, critical value ± 0.632.

Reject H_0, significant evidence of association between level and hardness.

2 (a) 0.787;

(b) $H_0 \; \rho = 0$, $H_1 \; \rho > 0$, test statistic 0.787, $n = 14$, 1 tail 5%, critical value 0.4575.

Reject H_0, significant evidence of positive association between body mass and heart mass.

3 (a) 0.784;

(b) $H_0 \; \rho = 0$, $H_1 \; \rho \neq 0$, test statistic 0.784, $n = 10$, 2 tail 5%, critical value ± 0.632.

Reject H_0, significant evidence of association inflation and unemployment rate.

4 (a) 0.610;

(b) $H_0 \; \rho = 0$, $H_1 \; \rho \neq 0$, test statistic 0.610, $n = 15$, 2 tail 5%, critical value ± 0.514.

Reject H_0, significant evidence of association between items produced and average quality score.

5 (a)

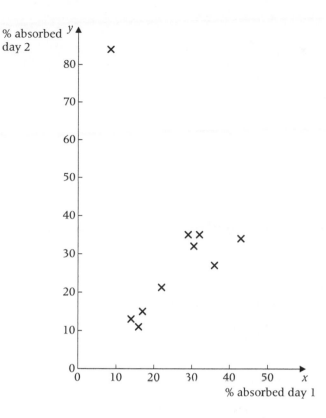

(b) -0.118;

(c) $\mathbf{H_0}\ \rho = 0$, $\mathbf{H_1}\ \rho \neq 0$, test statistic -0.118, $n = 10$, 2 tail 5%,
critical value ± 0.632.
Accept $\mathbf{H_0}$, no significant evidence of association between % absorbed
on first and second day.

(d) (i) patient 7,

     (ii) $\mathbf{H_0}\ \rho = 0$, $\mathbf{H_1}\ \rho \neq 0$, test statistic 0.863, $n = 9$, 2 tail 5%,
critical value ± 0.666.
Reject $\mathbf{H_0}$, significant evidence of association between %
absorbed on first and second day.

6 (a) -0.944;

(b) $\mathbf{H_0}\ \rho = 0$, $\mathbf{H_1}\ \rho \neq 0$, test statistic -0.944, $n = 15$, 2 tail 1%,
cv $= \pm 0.641$.
Reject $\mathbf{H_0}$, significant evidence of (inverse) association between
latitude and mid-temperature.

EXERCISE 10B

1 Conclude there is association between level of food supplement and shell
hardness when no such association exists.

2 Conclude there is no positive association between body mass and heart
mass when such an association does exist.

3 (a)

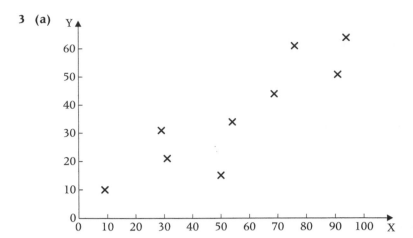

(b) 0.887;

(c) $H_0 \rho = 0$, $H_1 \rho > 0$, test statistic 0.887, $n = 9$, 1 tail 5%,
critical value 0.5822.
Reject H_0, significant evidence of positive association between X and Y;

(d) Type I error – conclude positive association exists when it does not.
Type II error – conclude no positive association exists when it does.

4 (a) 0.868;

(b) $H_0 \rho = 0$, $H_1 \rho \neq 0$, test statistic 0.868, $n = 8$, 2 tail 1%,
critical value ± 0.834.
Reject H_0, significant evidence of (positive) association between weight of ewes
and weight of lambs;

(c) Type I error – conclude association between weight of ewes and weight of lambs
exists when it does not.
Type II error – conclude association between weight of ewes and weight of lambs
does not exist when it does.

5 (a) 0.661;   **(b)** 0.995;   **(c)** 0.588.

(d) $H_0 \rho = 0$, $H_1 \rho \neq 0$, $n = 11$, 2 tail 5%, critical value ± 0.602,
ts 0.661, reject H_0, significant evidence of (positive) association between X and
Y,
ts 0.995, reject H_0, significant evidence of (positive) association between X and Z.
ts 0.588, accept H_0, no significant evidence of association between Y and Z;

(e) There is significant correlation between population and number with no
access to hot water supply and between population and number with
exclusive use of flush toilets. This has lead to positive (although non-
significant) association between number with no access to hot water supply
and number with exclusive use of flush toilets. The different populations of
the districts has upset the housing experts expectation of a negative
correlation. Correlation would almost certainly have been negative if
percentages had been used instead of numbers.

6 (a) (i) 0.908;   **(ii)** 0.593;

(b) $H_0 \rho = 0$, $H_1 \rho > 0$, $n = 10$, 1 tail 5%, critical value 0.549,
ts 0.908, reject H_0, significant evidence of positive association between IQ and Latin
score,
ts 0.593, reject H_0, significant evidence of positive association between IQ and
Music score;

(c) 1% significance level, cv 0.7155. Still significant evidence of positive association
between IQ and Latin but association between IQ and Music score not
significant at 1%.

EXERCISE 10C

1 0.750, fairly strong association between rank orders of preference and sweetness.

2 0.0952.

3 0.0500.

4 0.912, strong evidence that number of rotten peaches increases with number of days in storage.

5 **(a)** 0.922 (0.923 if $\sum d^2$ used);

   (b) Strong direct association between ranks of scores on the two questions.

6 **(a)** **(i)** 0.671 (0.673 if $\sum d^2$ used),
      (ii) 0.833,
      (iii) 0.958;

   (b) Student L has very poor examination result but good practical grade and satisfactory essay grade;

   (c) Essay rank is very closely associated with practical work grade. Hence little information will be lost if essay rank no longer recorded;

   (d) PMCC cannot be calculated for letter grades.

EXERCISE 10D

1 **(a)** H_0 pop ranks independent, H_1 pop ranks not independent, ts 0.0952, $n = 8$, cv 5%, two-tail test ± 0.7381.
      Accept H_0, no association between preference and price;

   (b) H_0 pop ranks independent, H_1 pop ranks positively associated, ts 0.750, $n = 7$, cv 1%, one-tail 0.8571.
      Accept H_0, no significant association between preference and sweetness;

   (c) H_0 pop ranks independent, H_1 pop ranks positively associated, ts 0.0500, $n = 9$, cv 1%, one-tail 0.7667.
      Accept H_0, no significant associations between colour and appearance;

   (d) H_0 pop ranks independent, H_1 pop ranks not independent, ts 0.912, $n = 10$, cv 1%, two-tail ± 0.7818.
      Reject H_0, significant association between days in storage and number of rotten peaches.

2 -0.600.
   H_0 pop ranks independent, H_1 pop ranks inversely associated, ts -0.600, $n = 5$, one-tail 5%, cv -0.900.
   Accept H_0, no significant association between judges' rankings.

3 0.444 (0.446 if $\sum d^2$ used).
   H_0 pop ranks independent, H_1 pop ranks not independent, $n = 9$, cv 5%, two-tail ± 0.6833.
   Accept H_0, no significant association between takings and % part-time staff.

4 0.595.
   H_0 pop ranks independent, H_1 pop ranks not independent, $n = 8$, cv 1%, two-tail ± 0.8571.
   Accept H_0, no significant association between reading age and rank of poem.

5 -0.798 (-0.783 if $\sum d^2$ used).
   H_0 pop ranks independent, H_1 pop ranks inversely associated, $n = 9$, cv 5%, one-tail -0.6000.
   Reject H_0, significant inverse association between rainfall and sunshine.

MIXED EXERCISE

1 (a)

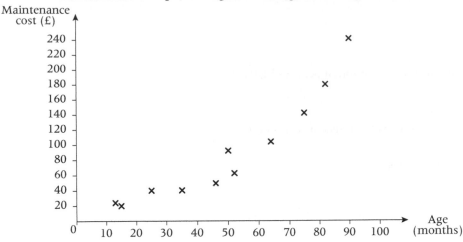

Maintenance cost plotted against the age of sewing machines

(b) 0.937;

(c) 0.973;

(d) H_0 pop ranks independent, H_1 pop ranks not independent,
ts 0.937, $n = 11$, two-tail 1%, cv $\pm$ 0.7348.
Reject H_0, significant evidence of association between maintenance
cost and age.
H_0 pop ranks independent, H_1 pop ranks not independent,
ts 0.973, $n = 11$, two-tail 1%, cv $\pm$ 0.7545.
Reject H_0, significant evidence of association between ranks of
maintenance cost and age.

2 (a) 0.745;

(b) H_0 $\rho = 0$, H_1 $\rho > 0$, ts 0.745, $n = 10$, one-tail 5%, cv 0.5494.
Reject H_0, significant evidence of positive association between IQ and
English score;

(c) -0.0952;

(d) H_0 pop ranks independent, H_1 pop ranks not independent,
ts -0.0952, $n = 8$, two-tail 5%, cv $\pm$ 0.7381.
Accept H_0, no significant evidence of association between ranks of
aptitude and perseverance;

(e) Students with high IQs tended to get high English scores, but
perseverance did not appear to be related to aptitude.

3 (a) 0.861 (0.862 if $\sum d^2$ method used);

(b) H_0 pop ranks independent, H_1 pop ranks positively associated,
ts 0.861, $n = 12$, one-tail 5%, cv 0.5035.
Reject H_0, significant evidence of positive association between ranks
of training and service score ranks;

(c) H_0 pop ranks independent, H_1 pop ranks not independent,
ts -0.808, $n = 12$, two-tail 5%, cv $\pm$ 0.5874.
Reject H_0, significant evidence of (inverse) association between ranks
of training and service score ranks.
The training scores awarded by the second manager are (strangely)
inversely associated with service scores. This manager appears to be
a poor judge of how trainees will perform later when working for
the company.

4 (a) Infant mortality

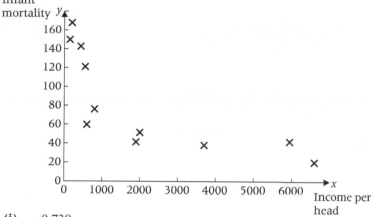

(b) (i) −0.739,

(ii) Scatter diagram shows relationship to be non-linear;

(c) −0.936;

(d) H_0 pop ranks independent, H_1 pop ranks not independent, ts −0.936, $n = 11$, two-tail 1%, cv ± 0.7545.
Reject H_0, evidence of significant association between ranks of infant mortality and income (association is inverse);

(e) Both correlation coefficients are negative but Spearman is numerically larger. This is because there is a strong but non-linear association between the variables as shown by the scatter diagram.

5 (a) (i) 0.899 (0.900 if $\sum d^2$ method used),

(ii) 0.794;

(b) H_0 pop ranks independent, H_1 pop ranks not independent, ts 0.899, $n = 10$, two-tail 1%, cv ± 0.7818.
Reject H_0, significant evidence of association between ranks of price and efficiency;

(c) Appearance is associated with price but on this evidence not as strongly as efficiency is associated with price. Hence the appearance ranking provides useful additional information. Suggest if amount of data is to be reduced 'efficiency' should be dropped rather than 'appearance';

(d) Not possible since efficiency grades are not numerical.

6 (a) Gross earnings (£000)

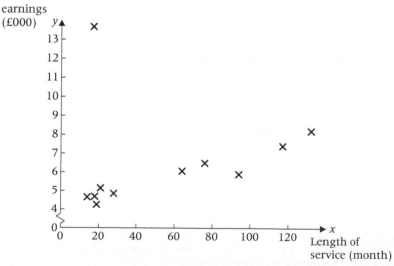

(b) 0.168;

(c) 0.518;

(d) $H_0\ \rho_s = 0$, $H_1\ \rho_s \neq 0$, ts 0.518, $n = 11$, two-tail 5%, cv ± 0.6091. Accept H_0, no significant evidence of association between ranks of length of service and gross earnings;

(e) Scatter diagram suggests that, apart from G who has short service and the highest earnings, there is a tendency for earnings to increase, approximately linearly, with length of service. Without the outlier (G) the correlation coefficients would be close to 1. The outlier has reduced the PMCC close to zero. It has also reduced Spearman's rank correlation coefficient but not to the same extent.

7 (a)

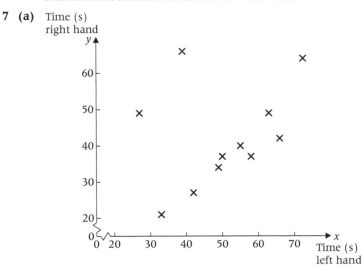

Time (s) right hand

Time (s) left hand

(b) 0.296; low positive correlation – insufficiently strong to suggest a linear relationship;

(c) $H_0\ \rho = 0$, $H_1\ \rho \neq 0$, ts 0.296, $n = 11$, two-tail 5%, cv ± 0.6091. Accept H_0, no significant evidence of association between times with left and right hands;

(d) E and G since they were quicker with their left than with their right hand.   0.954.
H_0 pop ranks independent, H_1 pop ranks not independent, ts 0.954, $n = 9$, two-tail 5%, cv ± 0.6833.
Reject H_0, evidence of significant association between times with left and right hands (association is positive);

(e) Strong positive linear relationship between times taken with left and right hands when left handers excluded.

8 (a)

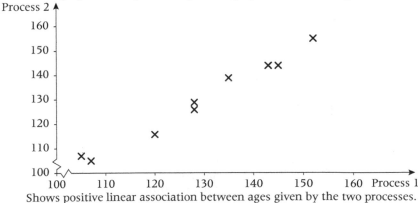

Age process 2 plotted against age process for 1 fragments

Process 2

Process 1

Shows positive linear association between ages given by the two processes.

(b) $r = 0.992$;

(c) $\mathbf{H_0}\ \rho = 0$, $\mathbf{H_1}\ \rho > 0$ (one-tail) 1%
cv $= 0.7155$ $r > 0.7155$
Reject $\mathbf{H_0}$. Significant evidence of direct association;

(d) $r_s = -0.498$ (or -0.494);

(e) $\mathbf{H_0}$ pop ranks independent $\mathbf{H_1}$ pop ranks not independent
(two-tail) 5%
cv $= \pm 0.6485$
$|r| < |cv|$ accept $\mathbf{H_0}$. No significant evidence to suggest any
association between age from process 1 and from colleague.
Colleague's claim not supported (assuming scientific processes are
reasonably accurate).

9 (a) Road distance

(b) F straight line distance exceeds road distance. Impossible – definitely
an error. H road distance much longer than straight line distance.
Unlikely but possible – probably an error;

(c) 0.406.
$\mathbf{H_0}$ pop ranks independent, $\mathbf{H_1}$ pop ranks not independent, ts 0.406
(or -0.406 if shortest distance ranked 1), $n = 10$, two-tail 5%,
cv 1 ± 0.6485.
Accept $\mathbf{H_0}$, no significant evidence of association between preference
and ranked road distance.

10 (a) $r_s = -0.754$ (or -0.748).

(b) $\mathbf{H_0}$ pop ranks independent $\mathbf{H_1}$ pop ranks not independent
(two-tail) 5%
ts $= -0.754$ cv $= \pm 0.6485$
$|ts| > |cv|$ Significant evidence to reject $\mathbf{H_0}$ and conclude that a
(negative) association exists between assets and ratings;

(c) (i) $r = 0.968$,

(ii) $\mathbf{H_0}\ \rho = 0$ $\mathbf{H_1}\ \rho > 0$
(one-tail) 1%
ts $= 0.968$ cv $= 0.7155$
ts $> $ cv Significant evidence to reject $\mathbf{H_0}$ and conclude that a
direct association exists between assets and number of
repossessions.

11 (a)

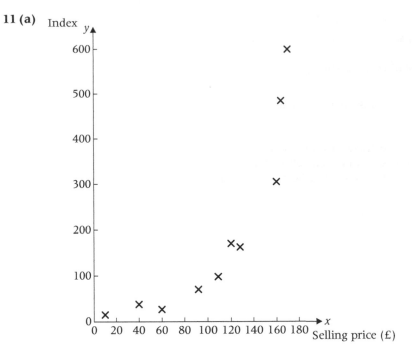

Index *y* axis, Selling price (£) *x* axis

(b) 0.976;

(c) Relationship appears to be non-linear;

(d) ts 0.976, $n = 10$.

  (i) H_0 pop ranks independent, H_1 pop ranks not independent, 10%, two-tail, cv $= \pm\,0.5636$, reject H_0, conclude selling price associated with index,

  (ii) H_0 pop ranks independent, H_1 pop ranks inversely associated, 1%, one-tail, cv $= 0.7333$, reject H_0, conclude selling price positively associated with index,

  (iii) H_0 pop ranks independent, H_1 pop ranks inversely associated, 5%, one-tail, cv $= -0.5636$, accept H_0, no evidence to conclude high selling price associated with low value of index;

(e) Alternative hypothesis determines that the critical value will be negative. Since test statistic is positive null hypothesis must be accepted.

Exam style practice paper

SS02

1

	1998				1999			
	Q1	Q2	Q3	Q4	Q1	Q2	Q3	Q4
Expenditure	163	265	409	186	165	293	469	196
Moving average			255.75	256.25	263.25	278.25	280.75	

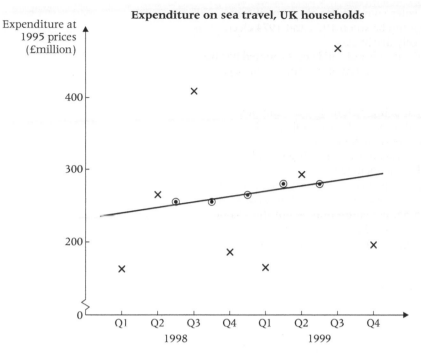

Expenditure on sea travel, UK households

(b) Trend suggests moving average of about 300 for Quarter 1 2000, Q1 1998 is about 75 below trend, Q1 1999 is about 105 below trend. $\frac{(75 + 105)}{2} = 90$. Forecast for Q1 2000 $300 - 90 = 210$

Forecast expenditure for Quarter 1 2000 is £210 million (it would be equally valid to base the estimate of seasonal effect on Q1 1999 only, which would lead to a forecast of about £195 million).

(c) The actual figure was well below the forecast. The forecast was made on a very small amount of data and not much should be read into the figures for one quarter but this does suggest that the upward trend may not be continuing.

2 (a) (i) 0.946, **(ii)** 0.113, **(iii)** 0.699;

(b) 3;

(c) (i) mean not constant – higher mean during the day than during the night,

(ii) calls not independent – cannot be answered if line engaged.

3 (a) $H_0 \ \mu = 60$, $H_1 \ \mu > 60$

$z = 2.96$, cv $= 1.6449$

Reject H_0, conclude mean time taken by applicants greater than mean for current employees;

(b) $z = 2.02$ cv and conclusion unchanged.

4 (a) (i) 1.44, **(ii)** 2.46, **(iii)** 0.620;

(b) 0.6;

(c) Regular commuter will nearly always carry sufficient coins. This computer probably needs a ticket costing less than £1.44.

5 (a) 30 671 000.

(b) (i) 165, **(ii)** 66 500.

(c) (i) The number of electors has shown an upward trend. Apart from the jump for the 1970 election due to the reduction of the minimum voting age, this has been a steady approximately linear increase.

    (ii) The number of valid votes counted has also shown some upward trend. If the jump between 1970 and 1974 elections is ignored the trend largely disappears.
        The upward trend of the number of valid votes counted has been less steep and more erratic than that of the number of electors.

6 (a) Systematic sampling;
   (b) Yes, all have probability $\frac{1}{12}$ of being included in the sample.
   (c) No, not all subsets of size 50 possible, for example two friends shopping together could not both be included.
   (d) No, first 12 customers have probability $\frac{1}{12}$ of being included, the next 88 have no chance of being included.
   (e) Likely to be representative. If customers entering later in the day had, as a group, different views than those entering early in the day the first sample would not be representative but the second would be. If customers coming on different days had, as a group different views then the samples would not be representative.

Exam style practice paper

SS03

1 (a)

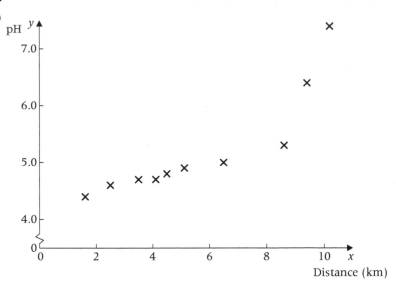

    pH starts well below the desirable level and increases approximately linearly with distance from about 2 km to 8 km. It then apparently increases at a much faster rate with distance from about 8.5 km to 10 km. The final observation is just above the desirable level;

   (b) 0.997;

   (c) H_0 pop ranks independent, H_1 pop ranks not independent
cv $= \pm 0.6485$. Significant association between ranks of distance and of pH values. Close to the factory the pH levels indicate acidity. They increase with distance and reach the desirable level at about 10 km from the factory. No data to indicate what happens beyond 10 km;

   (d) The scatter diagram shows a non-linear relationship although the ranks of the data are in almost complete agreement. Product moment correlation coefficient will be less than Spearman's rank because of the non-linearity.

2 **(a)** ts = 7.34   cv = 5.991
Conclude answer dependent on area;

(b) **(i)**

	Access to car	No access to car
More money on public services	60	90
Reduce petrol tax	85	15

(ii) 48.0   cv 3.841
Answer not independent of access to private car.

(c) A larger proportion of adults in area A wish to reduce petrol tax. A much larger proportion of adults with access to a private car wish to reduce petrol tax. The first result is primarily due to area A containing a greater proportion of adults with access to a private car.

3 **(a)** Experimental error reduced by both modules being judged by the same students. More likely to detect a difference if a difference exists;

(b) T = 12.5, cv = 14. Conclude there is a significant difference – module 2 higher evaluation;

(c) $P(\leqslant 2+) = 0.0193$. Conclude students have performed below expectation;

(d) Students appear to perform less well than average and so the result in **(b)** may not apply to a typical group of students.

4 **(a)** H − 7.21   cv 5.991
Significant evidence that not all populations are identical;

(b) The hypothesis test only establishes that not all the populations are identical, however the data suggest that percentage porosity decreases as temperature increases (at least in the 1400 °C–1500 °C region).

Index

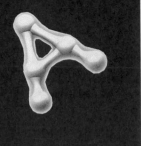